DICTIONARY OF HYDROLOGY
AND WATER RESOURCES

DICTIONARY OF HYDROLOGY AND WATER RESOURCES

2ND EDITION

GORDON STANGER

authorHOUSE®

AuthorHouse™ UK
1663 Liberty Drive
Bloomington, IN 47403 USA
www.authorhouse.co.uk
Phone: 0800.197.4150

© 2014 Gordon Stanger. All rights reserved.

No part of this book may be reproduced, stored in a retrieval system, or transmitted by any means without the written permission of the author.

Published by AuthorHouse 06/05/2018

ISBN: 978-1-4969-7717-5 (sc)
ISBN: 978-1-4969-7718-2 (hc)
ISBN: 978-1-4969-7719-9 (e)

Print information available on the last page.

Any people depicted in stock imagery provided by Thinkstock are models, and such images are being used for illustrative purposes only. Certain stock imagery © Thinkstock.

This book is printed on acid-free paper.

Because of the dynamic nature of the Internet, any web addresses or links contained in this book may have changed since publication and may no longer be valid. The views expressed in this work are solely those of the author and do not necessarily reflect the views of the publisher, and the publisher hereby disclaims any responsibility for them.

Cover photo:
Flooding in Bangkok, partly due to extreme monsoonal rains, and partly to land subsidence caused by hydrocompaction, and consequent local reversal of the surface hydraulic gradient.

TABLE OF CONTENTS

Preface to the Second Edition vii

- A. 1
- B. 27
- C. 44
- D. 85
- E. 112
- F. 132
- G. 149
- H. 165
- I. 183
- J. 200
- K. 201
- L. 204
- M. 214
- N. 234
- O. 240
- P. 248

Q	277
R	280
S	303
T	359
U	378
V	382
W	389
X	401
Y	402
Z	402
Appendix	421

Preface to the Second Edition

Much has changed since the first edition of 'DHWR' was published in 1994. The global per capita water availability has decreased by at least 22%. Groundwater mismanagement abounds, with an insane 'race for the bottom' in most of the world's major aquifers. Water pollution has changed in nature. Climate change denial notwithstanding, the world's water resources are noticeably changing, with wet areas getting wetter and dry areas getting drier. One estimate is that, by 2040, some four billion people, about half the global population, will be living in water-stressed regions. Sea level rise looks set to degrade most of the world's coastal aquifers - with the paradoxical exception of Greenland. In terms of water management there are some grounds for optimism such as a 14% improvement in the proportion of people served by sanitation, and improved monitoring of water quality, whilst our knowledge of hydrology and water resources has increased massively. Alas, one cannot say the same of our collective wisdom. Nowadays few water and wastewater utilities are still run by technocrats. They have mostly been high-jacked by lawyers, accountants and bureaucrats whose grasp of technical matters, as opposed to profit margins, is often woeful.

Naturally, the jargon evolves in response to all such changes becomes more management and less technically oriented. Exceptions to this 'dumbing down' trend lie in academia, where concepts involving fuzzy logic, decision support systems, and neural networks, which were never heard of 25 years ago, are becoming trendy.

As in the first edition, I have tried to simplify some of the obfuscation that can make this subject somewhat intimidating, both for the student and practitioner alike. Most water-related ideas are, after all, fairly simple in concept, though maybe not so much in practice. In a few definitions the reader may discern hints of some of my robust views, which have been fashioned over 45 years of practice in 30 countries and five continents. My perspective is a complex mix of amusement, cynicism, irritation, frustration and joy. But to the wise practitioner, the water sector is still one of the most rewarding careers available.

Gordon Stanger

This edition has deleted some of the obsolete words, is 32% larger, contains many expanded definitions, several improved definitions, and even a couple of minor corrections (whoops!) It is available in hard-back, soft-back and e-book. Best of all, it is bigger, better and, like the *Hitch-hikers Guide to the Galaxy*, cheaper! Enjoy.

Gordon Stanger, Hallett Cove, South Australia, June 2018
Comments and suggestions welcomed to <u>hydrodocgeo@yahoo.co.uk.</u>

Cover photo:
Flooding in Bangkok, partly due to extreme monsoonal rains, and partly to land subsidence caused by hydrocompaction, and consequent local reversal of the surface hydraulic gradient.

A

abandoned channel An ox-bow or billabong.

abandonment ... *of a well* ... Any such decommissioning should involve surface sealing of a water table aquifer to prevent contamination, or backfilling of an artesian aquifer to prevent loss of pressure.

ablation Loss of snow and/or ice by melting and evaporation. The total loss, in mm.day^{-1} is estimated as $M = C_m(T_{air} - T_m)$ where C_m is the ablation rate in mm.°C^{-1}.day^{-1}, T_{air} is the ambient daily temperature, and T_m is the threshold melt temperature. Naturally, the ablation rate is much greater in exposed snow-packs, than in forested areas.

abrasion Fluvial erosion by the scouring action of suspended or bottom sediments.

ABS (1) Alkyl benzene sulphonate. A typical anionic surfactant.
(2) Acrylonitrile butadiene styrene. A tough heat resistant thermo-plastic. One of several possible materials used to make well casing, usually in shallow aquifers.

absolute density = **Solids density**. The mean density of solid rock particles, excluding the pore space.

absolute humidity (p_w). Mass of water per unit volume of air at a fixed temperature. units of g.m^{-3}.

absolute porosity Linked porosity + non-linked porosity.

absorbance The modern parameter for analyzing and reporting water colour, as opposed to **colour units** *(q.v.)*. This is the ratio of absorbance to transmittance over a path-length of a metre, in units of Ab.m^{-1}. It is valid only for a specific wavelength and pH. For example, observe the decrease in absorbance when you add acid to tannin (lemon juice to tea). Most results are reported at the peak absorbance for natural organics, at 390 or 400 nm. Absorbance and colour are aesthetic parameters, which are, despite customer satisfaction, is unrelated to a water samples acceptability on health grounds.

absorption The water absorbed by any material (such as aquifer matrix) expressed as weight % of the dry weight. Absorption is not necessarily equivalent to the porosity.

absorbing well A borehole or dug well, used to recharge surface water to a confined aquifer. (Unconfined aquifers receive direct surface recharge).

abstraction = **groundwater extraction**, = **groundwater withdrawal**. The global annual groundwater abstraction amounts to about 1000 km^3, of which 80% is from only 15 countries. About 60% of global abstraction is used for irrigation.

Abyssinian well A primitive variety of borehole consisting of a drive point and slotted screen.

acicular ice Also known as satin ice. A form of freshwater ice which grows in a fibrous habit, often with entrained air bubbles.

acceptance The capacity of a water body to accumulate a pollutant before it reaches its maximum permissible concentration.

accepting aquifer Any aquifer with an active discharge zone, and hence with unhindered through-flow. Consequently, there is capacity for continuous recharge.

Dictionary of Hydrology and Water Resources

accretion *See* **collision.**

accumulated deficit/excess The cumulative deviation away from the mean, of variables such as temperature or precipitation. A reference time must be specified, such as the start of a month, year or decade.

accuracy A measure of the likely range of deviation from a true value. It is a relative measure, such as $M \pm x\,\%$, where M is the measurement and x is the accuracy. *cf.* '**precision**'.

acidity The milliequivalents of a strong alkali required to titrate a unit volume (usually a litre) of acidic water to neutrality.

acidization = Acidification. A potentially dangerous process of injecting acid into a well in order to: (1) locally increase the aquifer porosity by the dissolution of carbonate, or (2) to remove carbonate encrustation from a screen, pump or rising main. The process may improve the well efficiency, and increase the hydraulic conductivity of the aquifer in the vicinity of the well. Hydrochloric is usually the preferred acid because it is cheap and leaves non-toxic residues. Large quantities of acid may be required to make a significant improvement in the yield.

acid precipitation A more general term than acid rain, which includes acid fog, mist and dew. The latter categories leach acids from the atmosphere, but are less diluted than raindrops, and hence tend to be of lower pH. Urban fog tends to be more acidic than rural fog, due to sulphuric and nitric acid emissions. In extreme cases the resulting pH can be as low as 1.5.

acid rain Strictly, all rain is acidic because raindrops dissolve atmospheric CO_2 to become weak carbonic acid, with a pH of about 5.6. 'Acid rain' mainly refers to rain which has been further acidified by the absorption of SO_3^- and NO_x derived from human activities, especially the burning of fossil fuels. Although imprecisely defined, we may take the pH of most 'acid rain' as varying from 5.5 to about 3.5. Acid rain became a matter of growing concern in the 1960s and 1970s. It is still a significant issue in some parts of the world, although various national 'clean air acts' have reduced global acidification by between 40% and 65% relative to the 1975 baseline

acid sulphate soil   When saturated organic and sulphide-rich soils are drained, they oxidize, releasing sulphuric acid, iron and sometimes other toxic ions such as Al^{3+} and $As^{3+,5+}$. The resulting soil moisture pH drops to typically about 4.5 to 3.0, with dire consequences for any farm crops. One of several typical reactions is:
$$FeS_2 + 14Fe^{3+} + 8H_2O \rightarrow 15Fe^{2+} + 2SO_4^{2-} + 16H^+$$
Low-lying coastal areas, floodplains, swamps and tidal reaches of rivers are at particular risk. Examples are South Australian lakes, much of coastal Eastern Australia, and the western Mekong delta of Vietnam. The problem is treatable by liming, leaching, and careful control of the water table.

acidulous water   Sour (acidic) tasting water, especially associated with organic acids or high sulphate concentrations.

acoustic flowmeter   = Acoustic current meter. A means of measuring the mean stream velocity at a set depth by Doppler acceleration / retard-ation of a sound impulse obliquely across a stream. Complex to set up, and requires a stable uniform stream section, but has the advantage of not disturbing the flow regime.

acoustic log   See **sonic log**.

Dictionary of Hydrology and Water Resources

acoustic sounding   A sonic method of measuring the depth to water level in a borehole or stilling well.

acre-foot   A quaint and ancient unit of crop irrigation, still used in some backward parts of the world, such as the USA. 1 acre foot = 43560 ft^3 = 1230 m^3 = 1.23 Ml.

Actinomycetes   A globally common filamentous bacteria responsible for generating objectionable odours in stagnant water supplies. Although *Actinomycetes* is generally thought of as an aquatic bacterium, it is also common in soils.

activated carbon   = Activated charcoal. A highly porous granular substance (granular a.c. or 'GAC'). The carbon is obtained by the thermal anoxic carbonation of cellulose, and it is activated by treatment with chemicals such as $ZnCl_2$, after which it is calcined and washed. This creates a huge surface area (500 to 1500 m^2.g^{-1}) which gives the carbon an enormous adsorption capacity, whence its use in water purification. The source material (wood, coal, peat, coconut shells *etc.*) produces carbons of differing absorbtion capacity. A.C. is used to adsorb halogens and a wide variety of organic pollutants.

active glacier   A glacier in which some or all of the ice is moving. Many glaciers alternate between active and inactive periods. Many more have only recently become active in response to climate change.

active layer   = **Active zone** = **the supra-permafrost layer** = the layer of diurnal and seasonal freezing and thawing in tundra regions. Liquid water does not flow below the active layer, so this layer tends to be poorly drained, with bogs and seasonal ponds.

active protection   Extraction of groundwater before it flows into a mine working. In addition to minimizing water flow into the mine, part of the objective is to avoid contamination of the groundwater, and so drain the mine in an environmentally acceptable manner.

active storage   That reservoir volume consisting of the difference between maximum storage and dead storage.

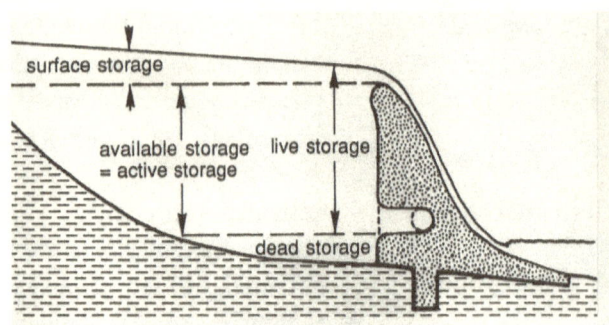

activity   (1) ... *of a clay* ... more correctly known as the *activity ratio*, $=\dfrac{I_p}{W_2}$ where I_p is the plasticity index (see Atterberg limits), and W_2 is that weight % of the soil/clay fraction which is < 2μm in particle size. Typical values range from about 0.02 to 7. Active clays are characterized by a high water-retaining capacity, swelling and shrinking, high CEC's, low permeability and possible thixotropy. Active clays have high cohesion, but deform easily under stress. It is mainly active clays such as attapulgite, smectite and some illites, which tend to cause engineering problems. The strength of low activity clays is derived from internal friction rather than from cohesion.

activity   (2) ... *of a solution* ... the effective or thermodynamic concentration, 'a', which is related to the molal ionic concentration, 'm', by the activity coefficient, 'γ', such that: a = γm. In most dilute solutions such as rainwater and some streamwaters, $\gamma \approx 1.0$.

actual evapotranspiration Evapotranspiration that actually takes place, as opposed to the potential evapotranspiration.

actual velocity (v_h) The rate of actual movement of bed-load particles, <u>not</u> including stationary intervals. *cf.* **virtual velocity**. A parameter used in fluvial sediment transport equations.

adaptor = **sub** = **coupling**. A connecting device for joining screen or casing of different thread or diameter.

adjusted stream Stream-flow along strike of the underlying geology.

adolescent ... state of a river. A well developed stream channel intermediate in form between the turbulent youthful state of the headwaters, and the sluggish meandering state of the flood plain (if present).

adsorption (= adsorbtion) The process by which certain ions in solution are attracted to and held by electrically charged particles. Some clay species, and Al and Fe oxy-hydroxide species are particularly noted for their strong adsorption properties. The process is strongly pH dependent, but in normal surface and groundwaters it is almost entirely cations that are affected. *cf.* **cation exchange capacity.**

advection The process by which a substance is laterally transported, such as atmospheric moisture swept by air currents across a catchment, or the percolation of a groundwater pollutant through an aquifer.

aeration Various techniques of preventing or minimizing the build-up of offensive reducing conditions in a reservoir.

aerobes Micro-organisms that require oxygen to metabolize carbon food sources. They may be subdivided into obligate aerobes, which must have dissolved oxygen, and facultative aerobes, which can grow in either an oxic or anoxic environments by reducing various oxidized inorganic solutes.

affluent = **fluvial tributary**.

afflux
(1) The difference in elevation between the undisturbed water surface, and 'backing up' due to channel constriction such as the piers of a bridge.
(2) The differential stage, under flood conditions, between immediately upstream and downstream of a weir.

Afflux can be complicated by choking, sediment transport, scour and débris accumulation.
In general, the equation used for computation of head-loss from a Flat Soffit Bridge constricting *lateral* flow is:

$$h_1^* = K^* \alpha_2 \frac{V_B^2}{2g} + \alpha_1 \left[\left(\frac{A_B}{A_4}\right)^2 - \left(\frac{A_B}{A_1}\right)^2 \right] \cdot \frac{V_B^2}{2g}$$

where h_1^* = the total backwater (or afflux) (m), K^* = total backwater coefficient (m), α_1 = kinetic energy coefficient at the upstream section, α_2 = kinetic energy coefficient in the constriction, V_B = average velocity in constriction (m/s), AB = gross water area in constriction (m²), A_4 = water area in the downstream section (m²), $A1$ = total water area in upstream section including that produced by the backwater (m²).

Dictionary of Hydrology and Water Resources

afforestation The planting and cultivation of new forest. The hydrological effect is to reduce catchment runoff by increased interception and evapotranspiration. Where temperate forest replaces grassland or shrubland the reduction in runoff is typically $44\pm3\%$, and $31\pm2\%$ respectively. In the case of Eucalypts the reduction is $\sim75\pm10\%$. Where the runoff is less than $\sim10\%$ of the rainfall, afforestation can 'mop up' *all* of the runoff. In the long term, afforestation by coniferous monoculture causes soil acidification. The older the afforestation, the more the runoff is reduced. Quite apart from the hydrologic loss, mono-cultured afforestation can be a form of environmental pollution.

afterbay The outlet reach of turbulent, and often fast-flowing water which emanates from hydroelectric turbines and other tail-races.

aggrading river A reach of a river which is actively depositing sediment.

aggregation *See* **collision.**

aggressive water A vague term for water that is chemically active in attacking concrete or metal. Especially acidic, strongly oxidizing or H_2S-rich waters.

airlifting Removal of water or aquifer material from a borehole by injecting high pressure air. In the case of shallow fine-grained aquifers, the drilling process itself can be achieved by air flushing (a very unpopular method with drilling crews), but air lifting normally refers to either developing the well, or to a crude form of well testing.

air line correction In velocity measurements from a cableway, there are two deviations from the vertical; above and below the water level. The air line correction determines the true distance from the cableway to the water surface.

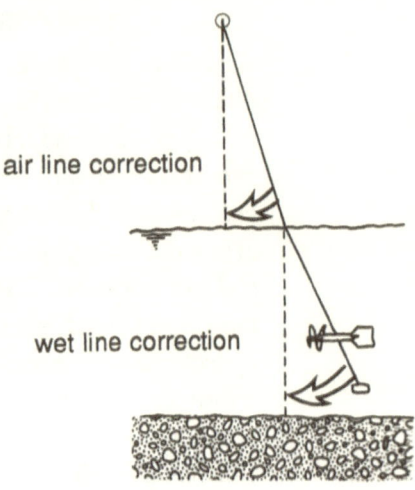

air stripping Removal of volatiles from water by means of injecting and venting air.

Akin triangles Division of a catchment into triangular areas, in which the apices are the location of raingauges, in order to estimate areal rainfall. The method is slightly more flexible and subjective than the conventional Thiessen polygon method.

Alternative Akin triangle constructions: The area influenced by the centre gauge is in bold

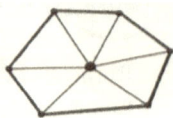 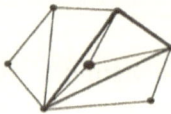

ALARA The acronym: 'As low as is reasonably achievable'. referring to the radiation exposure, especially for workers in the nuclear industry. It is a commonsense maxim which recognizes that zero exposure is absurdly impracticable.

alarm level A predetermined incipient flood level at which flood warning / evacuation measures are automatically triggered. Acceptable alarm levels tend to drift over time in response to improving storm-track monitoring, stage monitoring and climate change.

alas (Russian) A crater-like depression caused by localized thawing of permafrost, now associated with the breakdown of organic matter, and hence CH_4 release caused by global warming.

albedo The fraction of incoming radiation, averaged throughout the day, that is reflected back into the atmosphere. The albedo of a crop or land surface has an important bearing upon the amount of evaporation. Some typical values are:

fresh snow	0.8 to 0.9	shallow water	0.08 to 0.15
old clean snow	0.72 to 0.75	deep clear water	0.04 to 0.07
dirty snow	0.4 to 0.72	oceans	about 0.38
clouds	0.28 to 0.92	grassland	0.22 to 0.28
wet/dry soil	0.10 to 0.35	bare wet soil	about 0.11
dry light sand	0.20 to 0.6	deciduous forest	0.17 to 0.23
green cereals	0.20 to 0.26	coniferous forest	0.10 to 0.15
deserts	0.30 to 0.4	eucalypt forest	0.05 to 0.10
whole Earth	about 0.40	bare roc	0.2 to 0.6

albuminoid nitrogen A rather vague parameter for organically bound nitrogen in amino- and some other forms. The term is becoming, if not already, obsolete, although it was once widely quoted in water analyses. Albuminoid nitrogen ≠ **organic nitrogen**, *(q.v.)*.

alert stage = **caution stage**. The stage at which rising river levels should trigger the relevant authorities to take action, such as issuing a flood warning, or opening flood relief gates.

algal blooms Under ideal conditions of warm phosphate-rich water, cyanobacteria sometimes undergo dramatically rapid growth to produce 'algal blooms'. This commonly has a green soupy appearance, and may accumulate as a scum on river banks. The bloom colour may be grass-green, glaucous-green or even red. Some species, such as *Anabaena, Microcystis, Nodularia* and *Cylindrospermopsis*, are highly toxic, or even lethal to fish and livestock. In sea water, nitrate, rather than phosphate, is usually the limiting nutrient. There is no agreed consistent threshold concentration for defining a bloom, but concentrations are typically hundreds to thousands of cells per milliliter. With climate change and more intensive land use, algal blooms are becoming increasingly common and problematic in many countries.

algicide Many reservoirs, especially in hot countries, require treatment with an algicide, usually Cu-sulphate, to prevent the formation of algal blooms. Within 24 hours most of the copper is either precipitated or adsorbed onto suspended particulate matter. The dose rate is normally 1 to 2 $mg.l^{-1}$ as $CuSO_4.5H_2O$, (or about 0.25 to 0.50 $mg.l^{-1}$ as Cu; see 'copper'). Higher doses risk breaching public acceptability, whilst significantly lower doses risk some algae surviving the treatment and building up copper tolerance against subsequent treatments.

alkaline A solution on the high pH side of neutrality (under normal temperatures, neutrality occurs at pH 7.0).

alkalinity The ability of a solution to neutralize acids. Alkalinity is determined by titration against a strong acid. The 'end point' (at which ionic species contributing to the alkalinity are all neutralized) varies according to the solution, but is typically about pH 4.0 to 4.6, which may be indicated by pH electrode or methyl orange indicator. Note that it is possible for a solution to have *alkalinity*, and yet not be *alkaline*.

Dictionary of Hydrology and Water Resources

allogenic recharge ... or an 'allogenic system', is an aquifer in which recharge is indirect. Precipitation falls beyond an aquifer's immediate recharge zone, but is subsequently channelled into the aquifer by surface runoff.

allogenic stream Reaches of a stream whose flow is derived from distant head-waters.

alluvium A 'bucket term' for any sediment deposited by a river. The particle size can range from clay to huge boulders.

alpha radiation The least penetrating of the four major types of nuclear radiation, notably associated with uranium and transuranic decay series. It consists of energetic helium nuclei (2p + 2n) in which the large mass is easily stopped by heavy water, skin, paper, etc.

alternate depths For any given specific head, there are two alternative depths of flow, one supercritical and one subcritical, (except for flow at the critical depth).

alum Hydrated aluminium sulphate, commonly used as a defluoridating agent by precipitating Al-hydroxides. Complex polymerizing $Al(OH)_3$ then precipitates, providing an adsorption surface suitable for scavenging F^- ions from solution. Alum is also widely used as a flocculating agent in general water treatment.

aluminium Although the third most abundant element geochemically, Al normally occurs at sub-ppm concentrations in natural waters (except under acidic conditions). Al is highly toxic to fish and a few other aquatic organisms, and is toxic to humans at high concentrations. A possible link between Alzheimer's disease and high aluminium concentrations in drinking water is under investigation, although the evidence for this link is equivocal. Typical daily per capita ingestions are: from air 20 µg, from water 250 µg, and from food 7000 µg. However, the bio-availability of Al from each source is unknown. Different authorities set the MPL in drinking water at 50 and 200 µg.l^{-1} of Al. It is difficult to analyze aqueous Al accurately because some ionic species tend to polymerize.

alveolar A highly porous honeycomb texture, as found in the weathering and dissolution of some rocks, such as peridotite, which sometimes weathers to leave a residual quartz-goethite mesh texture.

ambient conditions The normal surface conditions of light, humidity, temperature, atmospheric pressure, *etc*.

American Petroleum Institute Although not directly connected with water, the API defines units and supplies standards and information which are used in hydrochemistry and in water well logging.

amictic Literally, 'without mixing'. Amictic lakes are alpine or polar lakes, with permanent stratification and ice cover, which are inherently stable, and never overturn.

ammonia NH_3. One of the less common, but highly soluble nitrogen species, formed by bacterial reduction of nitrate or nitrite, or by decomposition of nitrogenous organic matter. It is found in reduced groundwaters, but seldom occurs in surface waters.

ammoniation = Chloramination.

ammonium The cation NH_4^+. Most natural waters have a pH of < 9.2, in which case ammonium would dominate over dissolved ammonia.

Dictionary of Hydrology and Water Resources

amoebae   A group of unicellular protozoans, fairly common in natural waters and soils, where they play an important role in controlling fungi and bacteria. Different species occupy thermal niches from Antarctic lakes to thermal springs of > 42°C. Most are innocuous, but some are opportunistic pathogens, and one (*Naegleria fowleri*) is lethal.

anabatic wind   A convectionally driven warm wind which blows upslope from valleys. The rising cumulus towers arising from anabatic uplift are often locally important rain-producing mechanisms. Anabatic winds are much less common than katabatic winds.

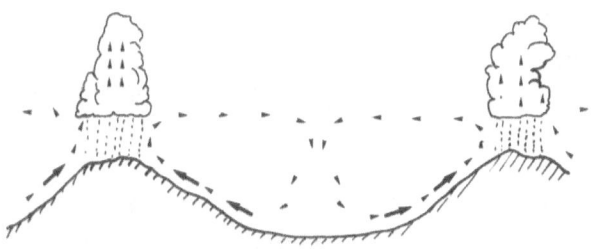

anabranch   A secondary river channel, often sub-parallel to the main channel, that carries active flow only during flood conditions, if at all.

anaclinal stream   A stream whose gradient is in the opposing sense to the dip of the underlying strata: a feature usually associated with antecedent or superimposed drainage.

analytical error   Referring to a water analysis: a measure of parity between cations and anions, defined as:

$$e = \frac{\left(\sum cations - \sum anions\right)}{\left(\sum cations + \sum anions\right)} \cdot 100\%.$$ An error of > 5% shows that there is something significantly wrong with the analysis. An error close to zero is ideal, but still not a guarantee that the analysis is correct. (For

example, an identical under- or over-estimate of both cations and anions might yield an error of 0%).

anastomosing channel *See* **braided channel.**

anchor ice = **bottom ice**. Ice, other than surface ice, which is frozen onto the bottom of a stream bed.

anemometer A device for measuring wind speed and/or wind run. Cup anemometers, consisting of three hemispheres mounted on a spindle, are most common, but sonic and other mechanisms are used for some specialist purposes. The data so collected is essential for estimations of evaporation.

angels Weather radar reflections from apparently clear air, due to abrupt changes of refractive index, or distant flocks of birds.

Angels' share Unaccounted for loss of water in a piped water supply, mainly caused by leakage, theft and poorly calibrated flow-meters.

anion exchange The process of exchanging negatively charged ions in solution with other negatively charged ions on a suitable substrate (such as an ion exchange resin). Anion analysis by ion chromatography is based on differential rates of anion exchange.

anisotropy In general, any variable whose point value varies according to direction. Specifically, the direction of groundwater flow is often strongly influenced by anisotropic hydraulic conductivity ($K_x \neq K_y \neq K_z$).

angle of draw When an underground void, such as a karstic dissolution channel, collapses, the subsiding overburden tends to assume the shape of an inverted cone. The angle of the conic surface, as measured from the vertical, is the angle of draw.

angularity correction If a current meter is inadvertently set at a slight angle (θ) to the flowlines, the apparent velocity can be adjusted to the true velocity by the angularity correction: $v_t = \dfrac{v_a}{\cos\theta}$

annual exceedance probability 'Self explanatory'; the probability of any given value of a hydrological parameter (such as a flood peak) being exceeded during a year. Estimates of the AEP are usually imprecise, but it is nevertheless a widely used statistical parameter.

annual flood The maximum discharge rate in any given calendar year or water year.

annual maximum series A data series consisting of the maximum value of each complete year of record. The mean of an annual maximum series will be less than or equal to the mean of an annual exceedence series.

annual series ... or annual exceedence series. A special case of a 'peaks over threshold' series in which the threshold is adjusted such that the number of peaks exceeding the threshold equals the number of years of record. (Not to be confused with the 'annual maximum series').

anoxic Without dissolved oxygen, (as distinct from 'hypoxic', *q.v.*). Reducing conditions are typical of deep hypolimnia, where decaying organic matter has 'mopped up' all of the available oxygen. Anoxic waters often stink of hydrogen sulphide, and contain anaerobic microorganisms. It is common for deep waters to be anoxic, even when the surface waters are oxygen saturated.

antecedence (1) An antecedent river channel is one which predates a period of geological uplift, and which therefore crosscuts the geology in apparent contradiction of the expected drainage direction. This antecedence should not be confused with the effects of river capture. Classic examples are the Indus and Brahmaputra rivers which crosscut the Himalayas.
(2) Antecedent weather conditions in a catchment may strongly influence the catchment hydrology. For example, x mm of rain may not induce any runoff at all after dry antecedent conditions, whereas a high proportion of the same rainfall may runoff after wet antecedent conditions.

Antecedent Precipitation Index (API). An estimate of the catchment soil moisture status which takes account of recent rainfall. The API $= kP_{n-1} + k^2P_{n-2} + k^3P_{n-3} + ...k^iP_{n-i}$, where n is the day number for the required API, P is that day's precipitation, and k is an empirical constant. The API is of questionable validity, and is certainly inferior to an instrumental water balance analysis, but the method provides plausible runoff estimates in statistically well-calibrated basins.

antidune migration Large sedimentary dune-like ripples that migrate upstream in fast moving water (Fr > 1). This kind of sedimentary motion is easily recognized by surface standing waves which closely follow the streambed contours, and which become steeper on the upstream side until they break.

AOP Advanced oxidation processes. A variety of techniques for producing OH$^-$ ions in water. Generating hydroxyl ions is an environmentally friendly way of getting rid of colour, COD, and many other organic compounds.

APDC	Ammonium pyrolidene dithiocarbamate. A 'wide spectrum' complexing agent, especially useful in the concentration of transition metals prior to AAS or ICPAES analysis. The concentration involves complexation and extraction into a small volume of a suitable solvent, such as MIBK or chloroform.

aphotic zone	Literally 'without light'. The aphotic zone may either be very deep or turbid water that is totally without light, or may be notionally taken as the depth below which < 1% of the surface light penetrates.

API units	American Petroleum Institute units. An arbitrarily defined unit of γ-radiation in which 16.5 API $\equiv$ 1 μg Ra per tonne.

apoplast	Part of a plant's water system, comprising the xylem, cell walls and any other part of a plant outside the protoplasts, as opposed to the symplast.

apparent density	Dry soil or sediment mass per unit volume (dried at 105°C).

apparent resistivity	In either surface or borehole resistivity, for any given electrode spacing, the apparent resistivity is usually a composite value of true resistivity from several zones or layers at differing distances from the electrodes. It is seldom possible to accurately deconvolute the true resistivity from apparent resistivity values, but neither is it always necessary. The apparent resistivity is a useful parameter in its own right.

apparent thickness	The apparent thickness of thick beds in a borehole resistivity log is erroneous. For conductive beds, the apparent thickness is > real thickness; for resistive beds, the converse applies.

application factor A safety factor used in aquatic pollution, limiting acceptable concentration = lethal concentration x application factor. For aquatic organisms it is typically in the order of 0.1, but very much lower application factors may be used for public drinking water, or in the case of cumulative toxins such as organochlorines.

approach segment Hydrograph recession just prior to a resumption of quickflow.

approach velocity See **coefficient of approach velocity.**

apron
(1) The platform, usually of concrete, at the top of a dug well which should drain outwards to prevent spilled water from running back into the well. This is necessary to prevent contamination, and hence the spread of infections such as Guinea worm.
(2) A concrete or other reinforced area at the toe of a dam, spillway or weir, for resisting turbulent erosion.
(3) A flat area of mineral precipitate (usually travertine or tufa) around a spring.
(4) A fluvial or glacio-fluvial outwash deposit, or roughly flat sedimentary layer in a piedmont setting.

aquiclude A rock unit impervious to water, such as unweathered and un-fractured rocks like fresh granite and most metamorphic rocks.

aquifer A saturated rock formation or stratum (Latin *aqua* = water, *ferre* = to bear). So most sandstones, gravels, limestones *etc.*

aquifer test Determination of the hydraulic properties of an aquifer by stressing it (by water extraction or injection) and observing the hydraulic response.

The term 'pump test' is a frequent misnomer for 'aquifer test'.

aquifuge A rock unit or stratum containing but not transmitting water, such as clay or shale.

aquitard A rock unit that transmits water, but very sluggishly, such as clayey silt.

Archie's law An approximation, used in borehole resistivity work: $R_{rock} = R_w \cdot a \cdot \phi^{-m}$, where R_{rock} is the bulk rock resistivity, R_w is the resistivity of the formation water, ϕ is the porosity, m is an empirical cementation factor, usually between 1.3 and 2.2, and a is a 'lithology factor' of between 0.6 and 2.0.

Archimedean screw A primitive pump consisting of a broad diameter helical screw closely set within an inclined tube. It is an ancient mechanism of low efficiency, but was one of the first advances upon simple bailing. Its use is limited to lifts of about a metre.

area-capacity The graph of elevation *vs.* surface area of a reservoir. The same graph often has a second scaled ordinate to indicate area and volume *vs.* height. This is taken as the total volume unless specified otherwise, (as in elevation *vs.* active volume).

areal reduction factor For a storm event of given intensity and duration over a given catchment area, the areal reduction factor is the ratio of the maximum *areal* precipitation to maximum *point* precipitation.

arenaceous A general descriptive term for all coarse to medium grained clastic particulate rocks, such as sand, sandstone, siltstone or conglomerate. Arenites generally behave as aquifers.

aridity Many attempts have been made to objectively quantify the dryness of a climate or area, including: the *Aridity index* (De Mortonne, 1926), the *Dryness index* (Budyko, 1956), and the *Moisture index* (Thornthwaite and Mather, 1955; Bailey, 1979). These employ functions of precipitation, temperature and potential evapotranspiration, but are only applicable on a regional scale. At smaller scales, the local variations in bare rock / soil cover, perennial biomass, rainfall intensity, and the extent of sand (giving rapid infiltration) are at least as important factors.

aridity index This is defined as $A = \dfrac{P}{T+10}$ where A is the index, P is the mean annual precipitation (mm), and T is the mean annual temperature (°C). It is not a good index for cold deserts.

argillaceous Characteristic of a fine-grained clay-rich rock, such as mudstone, shale or marl. Argillites behave as aquitards or aquicludes.

arheic Descriptive of any area devoid of surface drainage, such as the Nullarbor plain of South-West Australia.

armour
(1) hard-wearing protection around such items as borehole logging cables.
(2) In a stream, the near-surface fine-grained sediments may be winnowed away by the current, thus leaving a layer of coarse-grained sediment on the stream bed. This layer of 'armour' is stable, and inhibits further erosion. Armouring can ruin the applicability of many empirical bed-load equations. In fact, the physics of erosion, deposition and armouring is complex, and not amenable to simple analysis.

Dictionary of Hydrology and Water Resources

arroyo A wadi, creek or narrow valley, normally dry, and of sub-rectangular section. (Not the V section of a gully).

arsenic (As): A toxic metalloid whose aqueous concentration is usually below the typical threshold of detection of about 5 $\mu g.l^{-1}$. However, some parts of Utah and western USA, Alaska, Taiwan, Chile, Mexico, Vietnam, southern China, and all the sub-Himalayan and Mekong regions, have much higher aqueous concentrations, commonly between 50 and 500 $\mu g.l^{-1}$ (*cf* the WHO recommended limit of 50 $\mu g.l^{-1}$). In such countries the main risk is that of long-term exposure to As, resulting in skin cancer and other chronic conditions, rather than acute As poisoning. Some degree of arsenic poisoning is thought to affect up to 140 million people in 70 countries. Of the many possible aqueous species, $HAsO_4^{2-}$ is the most common (dominant between pH 7 and 11).

artesian An aquifer in which the water surface is bounded by an over-lying impervious rock formation. Here the water surface is at greater than atmospheric pressure, and so rises in any well or borehole which penetrates the overlying confining layer.

artesian well A well in which the water level rises until it reaches equilibrium (at atmospheric pressure). If the atmospheric pressure surface of the aquifer is above ground level, then the well is said to be 'artesian overflowing'. The static water level in an artesian non-overflowing well will equilibrate somewhere within the confining layer, or within a higher overlying formation.

artificial recharge Or: Induced recharge, artificial infiltration, or injection. Processes of diverting surface water resources into ground-water storage for later use. This may involve flood spreading, retention dams and/

or injection wells. Artificial recharge is particularly useful in arid zones, where surface storage would incur unacceptable evaporative losses. Groundwater storage is also less susceptible to biological and chemical pollution.

aspect The direction of a hill-slope with respect to the prevailing wind or flood-wave. Aspect is a significant physiographic parameter in mountain rainfall. The northern hemisphere convention is 'eastward deviation from the south is positive, whilst westward deviation from the south is negative'. In the southern hemisphere the sense of deviations is the same, but is measured with respect to a northerly line.

assimilation capacity The amount of pollution (undefined) that a site can absorb and contain.

athalassic *(A = not; thalassa = sea)* Descriptive of a non-marine origin despite similarities in salt content to that of sea water. Examples include some of the near-coastal lakes in Antarctica which are saline but of a distinctive geochemical origin.

atmometer An instrument which measures rates of evaporation from a moist filter paper or porous plate under standard conditions.

atmospheric pressure surface = the piezometric surface: the actual surface in unconfined conditions; a hypothetical surface under confined conditions.

Atomic Mass Unit 1 AMU = one twelfth of the mass of a ^{12}C atom ≈ the mass of a proton or neutron.

attapulgite An uncommon variety of clay used to make drilling mud under salt water conditions.

Dictionary of Hydrology and Water Resources

attenuation Apart from the general sense, the strict sense is the rate of reduction of a pollutant along a flow-path, *e.g.* % per 100 m.

Atterburg limits The plastic properties of a clay or soil, expressed in terms of their water content, which have long been used in foundation engineering. They are the plastic limit, P_w, which is a measure of the water content that can be held in the clay with some degree of rigidity; the liquid limit, L_w, which is a measure of the water content that can be held before the soil loses all rigidity and starts to flow; and the 'plasticity index':

$$I_p = L_w - P_w$$

authigenic Formed *in situ*. For example, directly precipitated clays, as opposed to clays of alteration, erosion or advective origin.

autocatalysis A reaction which is catalysed by the reaction product. Hence the reaction is sluggish to get started, but once initiated, the reaction speed accelerates to completion. A self-catalysing reaction.

autochthonous groundwater Groundwater derived from local recharge above the aquifer, as opposed to inflow from some laterally adjacent source. Especially associated with karstic aquifers.

autocorrelation A means of testing a time-series of data (streamflow, evapo-transpiration, precipitation, *etc*) for periodicity. The series or partial series is correlated with itself at differing time lags. Periodicity shows up as a peak in the auto-correlation-lag graph. As with all such statistics, the results should be weighed against a significance test.

autogenic Precipitation which recharges by vertical infiltration, either directly into the rock, or through a pervious soil. Especially applicable to direct recharge of karst aquifers. The alternative to autogenic recharge is allogenic recharge.

auxiliary curve A graph relating outflow from a reach to the storage-outflow parameter: $\left(\dfrac{S}{\Delta t} + \dfrac{O}{2}\right)$. This is obtained from field survey measurements, and is necessary for the solution of flood routing computations.

available soil moisture The difference in water content of a unit volume of soil between field capacity and wilting point.

average recurrence interval Synonymous with, but preferable to, the older term: 'return period'. The ARI is a statistical measure of the frequency of some event such as a storm, flood or drought. Inclusion of the word 'average', is very important: The occurrence of an X year flood this year does *not* mean that we have will necessarily have to wait another (X-1) years before the next flood of similar magnitude; rather that *on long-term average* we will expect such a storm once every X years.

The range of ARI chosen for design purposes varies greatly according to country, social and infrastructure values. Some plausible values (not to be taken too literally), might be:

purpose	range of ARI (yrs) for the design flood
major dam spillways	50 to the PMF
floodways, tunnels, bypasses	20 to 100
major bridges and culverts	50 to 100
minor bridges and culverts	20 to 50
high value urban structures	10 to 100

avulsion The process in which a river changes course to create a short-cut. That is, a new channel with a steeper hydraulic gradient. This is the cause of new distributaries evolving in a delta. Also, as meanders evolve, they eventually become so sinuous that avulsion occurs, with break-through and short-circuiting to leave an ox-bow or billabong.

AWS Automatic or Automated Weather Station. A weatherproof multi-channel data logger with a variety of sensors attached to a two to three metre mast. Measured parameters typically include air and soil temperatures, wind run and direction, rainfall, relative humidity, nett radiation and solar radiation.

backfill Replacing excavated material, or the material used in backfilling. It is especially important that abandoned dug wells and bore-holes are always backfilled. This is for safety reasons, to avoid contamination of the aquifer, and to maintain the hydraulic pressure of confined aquifers.

backflow In a pipe or channel, the backing up of water against the normal direction of flow.

backslope = **dipslope**. The low angle land slope away from a scarp.

backslope drainage = **consequent drainage**.

backwash(ing) = **backblowing**. A means of unclogging a filter:
(1) in a borehole by suddenly reversing the flow to dislodge bridging fines in the filter pack and formation, and
(2) in the filter bed of a water treatment plant.

backwater correction Used where stage is not uniquely proportional to discharge. Where this occurs, use two stage measurements: mean stage plotted against $\dfrac{Q}{\sqrt{dy}}$ should give a stable (albeit non-linear) relationship. (dy is the differential stage).

backwater curve A longitudinal stream profile upstream from a constriction.

bacteria Bacteria account for roughly a third of the primary biological activity in most aquatic environments. All natural water bodies, whether fresh, brackish or saline, are inhabited by both single and colonial bacteria, most of which are harmless to humans. There is no significant correlation between total bacteria and incidence of disease, although there is a strong correlation between *E.coli* and disease. In open water bodies, bacteria occur in planktonic form or on substrates. Planktonic bacteria can be free floating, adsorbed onto particulates, or in symbiosis with zooplankton, typically with 10^7 to 10^{11} bacteria per host micro-organism. Generally, much higher concentrations of aquatic bacteria also exist on epiphytic, epilithic or sedimentary substrates. Aquatic bacteria also exist as biofilms on pipe and tank surfaces, and as more specialized bacteria adhering to moist plant roots.

bacteria, continued In oligotrophic systems the bacterial growth efficiency can be as low as 0.01, whereas in most eutrophic systems it is about 0.5. In warmer waters, with high nutrient levels, the rates of bacterial growth are limited by chemical reactions, solar radiation, viral lysis and predation. The optimum temperature for different species varies from about 8°C to 45°C. Some outbreaks of toxic cyanobacteria occur in reservoirs during abnormally hot summers, a trend which is likely to increase with global warming.

badlands Barren ground, heavily dissected by ephemeral fluvial erosion. Some occur naturally, but most of the world's badlands result from overgrazing and other forms of land misuse.

baffle A deliberate obstruction to fluid flow for such purposes as: efficient mixing, prevention of scour, and dispersion of kinetic energy.

bahr Arabic word for sea, river or seasonal wetland. Examples include Bahr el Jabal, Bahr el Ghazal.

bailed sample A water sample which has been bailed from a well is likely to have been exposed to the atmosphere for a significant interval, and hence is inferior to a pumped sample for determining the hydrochemistry of the undisturbed aquifer.

bailed well A well, usually dug rather than drilled, in which the lift is by bucket.

bailer A tubular device used in boreholes to remove either cuttings or water.

bail test Used to determine the approximate hydraulic conductivity of a low-yielding (low K) well.

bai-u The heavy rain season of Japan. (Chinese equivalent = maiyu).

balling up When drilling hard rock, it is possible for the bit to roll on ball-shaped rock chips. This slows down operations because the bit has difficulty 'biting' into the rock face. This 'balling up' is solved by 'hammering'.

balneology — Relating to bathing, as in the supposed therapeutic value of bathing in many thermal or mineral spring waters. Well might you confuse balneology with baloney!

bank — A river bank is known as 'right bank' or 'left bank' as viewed downstream in the normal direction of flow.

bankfull discharge = Discharge at 'in-bank-capacity'.

bankfull width — The maximum width of channel flow before the channel is overtopped. The width at 'bankfull discharge'.

bank revetment — The lining of a channel, above and below the water level, usually with concrete or stone, to prevent erosion. Used along meanders, or lakes prone to severe seiching.

bank storage — Water which percolates laterally from a river in flood, some of which may flow back into the river during low-flow conditions. Bank storage may be an important source of aquifer recharge.

bank toe zone — A river section below the water surface under normal, or base-flow, conditions.

banquette — A horizontal step in the landward side of a levee to lengthen the seepage flowpaths and maintain stability.

bar — (1) Area of sediment deposition in the shallows of a channel. A hazard to navigation.
(2) obsolete unit of pressure, *cf.* **hectopascal.**

barite — $BaSO_4$, a dense mineral used as an additive to drilling fluids to increase the fluid density, and hence the stability of the open hole.

Dictionary of Hydrology and Water Resources

barium	A toxic minor to trace element in most natural waters. The mean dissolved concentration is about 45 μg.l^{-1} as Ba^{2+} (the HPL = 1 mg.l^{-1}). Its chemistry is similar to calcium, and it proxies for Ca in many minerals. The only significant Ba mineral is the highly insoluble barite, which effectively controls barium's solubility in most environments. Ba^{2+} is readily adsorbed onto Mn oxides.
barrage	A gated dam, generally fairly low, to maintain the pool level of a reach, or to prevent sea water ingress.
barrier boundary	An impervious obstruction to groundwater flow, generally lateral rather than vertical (a confining layer is not thought of as a barrier). Examples are sealed faults, dykes, cemented banks or fluvial terraces, and rapid transitions to pelitic facies.
barrier effect	The reduction in rainfall on the lower slopes of a mountain caused by loss of atmospheric moisture during ascent and descent of the moist air over an upwind 'barrier'.
barrier well	An injection or pumped well used to halt or intercept a plume of pollution.
base exchange	= **cation exchange**.
base flood	An American term for a 100-year flood. A nominal index for floodplain management. In this era of climate-change the concept of the 'base flood' is becoming increasingly unreliable and misleading.
baseflow	= **base runoff** = **dry weather flow** = **sustained flow** = **delayed flow**. Stream flow derived from medium to long term groundwater storage. There is no clear and definitive way to separate the components of baseflow and quickflow from a stream hydro-graph.

baseflow index The hydrograph area below the baseflow separation, as a fraction of the total baseflow area.

baseflow separation The notional separation of the surface runoff from the baseflow components of a stream hydrograph. Although conceptually simple, objective baseflow separation proves to be inordinately difficult in practice. Consequently, most separation methods are somewhat arbitrary.

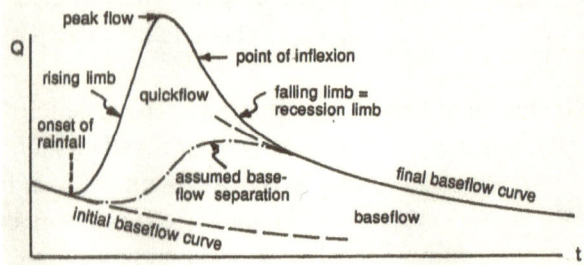

base level of erosion In any basin, the lowest bedrock threshold around the perimeter. Sedimentation, erosion and groundwater storage within the basin are all strongly controlled by the base level. In tectonically active areas, the base level of erosion may be smoothly or episodically time-variant.

base level of flow A stream profile at its transition from erosion to deposition.

basement *(geol.)* In any geological section, the first appearance of igneous plutonic or metamorphic rocks. This may lie anywhere from the surface to great depth.
(hydrogeol.) The impervious base underlying the deepest aquifer of a geological section.

base width The notional duration of a unit hydrograph.

basic rocks A class of silicate rocks, many of which are volcanic, which usually consist of Ca (± Na) feldspars +

Dictionary of Hydrology and Water Resources

pyroxenes or amphiboles, but *not* quartz. Examples include basalt and gabbro. The less common basic rocks include a wide variety of mineral assemblages. Apart from Hawaiian type basalts, the basic rocks are seldom good aquifers. In fact basasltic dykes sometimes form useful hydraulic boundaries. The basic rocks characteristically weather to yield Mg, Ca and SiO_2-rich groundwaters.

basin = Catchment area. Any drainage system. There may be a hierarchy of basins within basins.

basin irrigation Irrigation by flooding channels surrounding small plots of land, usually rectangular in shape.

bathylimnion The deepest, and essentially isothermal zone of a deep lake. Approximately synonymous with hypolimnion.

Bayesian network Alias **Bays Nets** or **Belief Network** combines elements of computing, statistics and probability to develop a kind of graphical mathematical framework representing knowledge of an uncertain domain. It uses probabilistic dependencies relating known and unknown variables. The originator of this kind of fuzzy modelling was Thomas Bayes in 1763, but practical BN applications only emerged in the 1980/90s. Diverse applications include weather forecasting, manage-ment of pipeline networks, and data transmission.

bayou = **anabranch**. An ephemeral floodwater channel. (U.S.) Open channel through a swamp.

Becquerel The modern (SI) unit of radioactivity. 1 Bq = 1 disintegration. Per second $\approx 2.7 \times 10^{-11}$ curies.

bed (1) A rock stratum, (2) The bottom of a stream or river.

bed layer	The layer immediately above the stream bed in which the bed-load is transported. It is nominally 'defined' as 'twice the sediment grain diameter', but this is vague: maximum, median, mean, or some other grain diameter?
bedload	= Traction load. Stream sediment that moves by rolling, sliding or saltation along a stream bed, as distinct from suspended load.
bedrock	Solid rock, as opposed to uncemented or loosely-cemented alluvium.
bela	(Indian) A reef, sandbar or island in a river.
belt of fluctuation	That part of an unconfined aquifer which undergoes periods of both saturation and undersaturation.
bench	= **river terrace**. A former river bank, now raised to a higher level. In some valleys, a succession of benches may be discerned, with the highest bench being the oldest.
benchmark	A mark set in stone or metal, of supposedly fixed elevation relative to mean sea level, and to which stage and groundwater elevations are referred. Unfortunately, changes in long-term sea level, and tectonic uplift/subsidence, may result in long-term errors in elevation.
bentonite	A variety of Na-montmorillonite clay, much used as a rotary drilling mud. It can absorb a huge amount of water to form a thixotropic slurry.
Bentzel tube	An instrument for measuring small point velocities in shallow streamflow.

Bergeron-Findeison The Bergeron-Findeison process is, perhaps, the most important of the raindrop-forming processes. Supercooled droplets, lofted by convection into the upper, cooler parts of a cloud, lose vapour which condenses onto ice crystals. Thus the ice crystals grow at the expense of the droplets because of the differential vapour pressure between the two phases. The enlarged ice crystals then fall, and melt into raindrops.

berm ditch = **side ditch**, = **counter drain**.

Bernoulli equation The basic energy equation of hydrodynamics in which energy, excluding friction, is expressed in head (metres): $\frac{v^2}{2g} + Z + \frac{P}{\rho g} = C$, or the total of kinetic, potential and pressure components are constant.

beta radiation High-energy electrons emitted from a nucleus by the decay of a proton to a neutron. Its penetration may be up to 2 cm into flesh.

billabong = oxbow lake. A river meander which has been cut off from the main channel by the normal processes of channel migration in a mature river.

biochemical oxygen demand (BOD) The oxygen required to oxidize any reduced species in solution. An important measure of water pollution.

bioengineering The emerging trendy practice of minimizing bank and slope erosion, gullying, and perhaps piping, by stabilizing the soil surface with appropriate vegetation. That is, by root binding, and reducing the surface runoff. This has been practised for decades, particularly on earth dam back-slopes, but until recently, it didn't have a fancy name.

biological assimilation The absorbtion of pollutants within organisms, for example, mercury in fatty tissue, or lead in bones.

bioremediation A growing and lucrative business of using micro-organisms to 'clean up' contaminated surfaces or groundwaters. There is great potential for genetically engineered organisms to target specific pollutants. However, there are also technical problems such as the availability of carbon 'food', oxygen supply, acclimatization, and sensitivity to temperature and toxicity.

bit The 'business end' of the drill string. There are many designs: chisel, star, roller cone, button, drag, etc. Naturally, the most appropriate bit is determined by the geology.

black-water stream (As distinct from 'black-water': sewage-contaminated waste-water). A stream or river characterized by a high concentration of organic acids, which colour the water red, brown or black. The classic black-water areas are parts of the Amazon basin where organic acids dominate the otherwise remarkably pure water chemistries.

bladder pump A low-discharge pump which utilizes pulses of compressed gas or air to sample water from wells.

blind zone On borehole resistivity logs a thin zone on either side of a high resistivity layer is 'electrically transparent'. Blind zones lie between the bed itself and its reflection peaks. Its width depends upon the dimensions of the bed and electrode array.

blood rain Rain coloured reddish-yellow by nucleating around iron oxide- rich dust. Dust from the Sahara, or Ethiopia, occasionally causes blood rain across Northern Africa and Europe, as in the account of the 'first plague', in Exodus 7. A blood rain incident in Kerala in 2001 was most likely caused by either toxic red algae (*Physteria*?) or bacteria, such as *Oscillatoria rubescens*.

bloom Discolouration of the water by vigorous growth of algae. Some blooms are toxic; all are deleterious to water quality due to excessive oxygen consumption and blocking of insolation.

blowout A surge of drilling fluid from a borehole as the bit enters a high pressure zone, such as 'puncturing' a confined aquifer. Blowouts can be spectacular, and a prelude to success, but can also be dangerous in the case of sulphurous gas release.

blue-green algae See **cyanobacteria**.

BOD Biochemical Oxygen Demand = Biological Oxygen Deficiency, This is the most commonly used index of water pollution, and consists of the amount of oxygen required by micro-organisms to oxidize all the reduced components (generally organic) in a litre of water. The analysis is undertaken over five days in a dark incubator at 20°C, and is conventional expressed as $BOD_5 = \ldots\ldots$ mg.l^{-1}. Typical values, in mg.l^{-1} are:

Very clear lake water	1 to 2
Moderately clean water	3 to 5
Somewhat polluted by organics	6 to 10
Water from logging and forest litter	20 to 80
Sewage effluent	200 to 400
Water from feedlots, cowsheds, etc	300 to 2000
Pulp and paper effluent	up to 20,000

Water quality standards often do not specify a BOD limit, but for most purposes one would expect a $BOD_5 < 3$ mg.l^{-1}.

bog Bog, swamp and marsh are often treated as synonymous, but strictly, a bog is a *lake shore* consisting of a spongy mass of vegetation, whereas a swamp is wet *land*. A bog is one of several possible end stages in the evolution of a lake. It is usually acidic and hypoxic.

boil In addition to thermal vaporization, a boil is a point eruption of water from a saturated sand. It may be as little as a few mm in diameter.

boiling point The temperature at which the total hydrostatic pressure equals the saturated vapour pressure. For a typical atmosphere, the boiling points of water vary with altitude (pressure):

altitude (km)	temperature °C	pressure HPa
0 (sea level)	100	1010
1	97	905
2	94	799
3	90.5	703
5	86	545

boisson A basin with radially inflowing drainage. The centre may be a lake, salt lake, or recharge zone to a karstic aquifer.

boom A floating chain or barrier (of many different designs) to contain oil spills or to protect a valve tower or other structure from floating débris or ice.

border ice Ice that preferentially forms along river banks. This is partly because of lesser water velocity, but mainly because shallow water cools more rapidly than deep water during cold nights.

border irrigation Flood irrigation between bounding ridges (borders).

bore A sudden change in stage caused by an advancing floodwave. Tidal bores travel upstream; flood bores may travel upstream or downstream. *See also* **borehole**.

borehole = **tubewell** = **bore**. Any drilled hole in the ground. Most hydrogeological data is obtained from boreholes. There is a rough distinction between *observation wells*—narrow boreholes used to measure water levels or to sample water, and *production wells*—wide boreholes whose primary purpose is to facilitate a pumped water supply. Borehole drilling is a substantial technology in its own right.

borehole geophysics = **Down hole geophysics**. A technology, largely developed by the oil industry, which yields information about a borehole. Normally a probe is lowered to the bottom of the hole immediately after drilling, whilst the hole is still filled with electrically conductive drilling fluid. It is then raised at a constant rate whilst various transducer signals are monitored and graphed. TV and temperature probes are logged during descent to minimize disturbance, whilst TV, temperature and sonic logs may be used long after drilling. Types of probes include: SP, long and short resistivity, density, γ, γ-γ, neutron, caliper and flow velocity.

bore volume The total volume of a borehole including the filter pack and any open annular space outside the casing.

boron (B). An essential trace element for plant growth, but toxic to plants at slightly higher concentrations. Permissible limits for irrigation waters are about 1 mg.l^{-1} for boron-sensitive crops, such as citrus trees, and 3 mg.l^{-1} for boron-tolerant crops, such as vegetables and palms.

bottom ice = **anchor ice** = **grounded ice**.

bottoms up time   In rotary drilling, the time it takes for the circulating drilling fluid to flush cuttings from the bit to the surface. It is only a few minutes in shallow water wells, but may be over an hour in a deep oil well.

Bouguer anomaly   The residual variation in measured gravity, due to variation in rock density, after all other factors have been taken into account. For example, a large negative Bouguer anomaly indicates low density, which might be accounted for by a thick high-porosity sedimentary basin, that is, a potential aquifer.

boundary condition Any computer model of surface or groundwater requires all of the boundaries to be specified as 'known head', 'known flow', or a mixture of the two. (*See* Dirichlet and Neumann boundaries). Lack of, or incorrect specification of, the boundary conditions will lead to either error or indeterminacy of the solution. Part of field hydrology therefore involves determining the hydraulic nature of a basin's boundaries.

bourne   An ephemeral streambed in chalk, nearly or entirely without baseflow (because it infiltrates into the porous rock).

Boussinesq equation $\frac{\partial^2 h}{\partial x^2} + \frac{\partial^2 h}{\partial y^2} = \frac{S}{T} \cdot \frac{\partial h}{\partial t}$ A linearized two-dimensional equation for horizontal groundwater flow associated with changing storage, where S is the storativity, and T is the transmissivity. If the aquifer is anisotropic then the equation may be written as:

$$T_x \frac{\partial^2 h}{\partial x^2} + T_y \frac{\partial^2 h}{\partial y^2} = S \cdot \frac{\partial h}{\partial t},$$ where T_x and T_y are the transmissivities in the x and y directions, respectively. A particular case of the Boussinesq equation, which

provides the basis for most aquifer test theory, is the radial flow equation: $\dfrac{\partial^2 h}{\partial r^2}+\dfrac{1}{r}\cdot\dfrac{\partial h}{\partial r}=\dfrac{S}{T}\cdot\dfrac{\partial h}{\partial t}$.

Bowen Ratio The ratio of sensible heat to latent heat, (β = H/LE). Since the sensible and latent heats are difficult to measure in the field, their ratio finds useful application in studies of evapo-transpiration.

brackish water Water of intermediate salt content between fresh and saline, for example, with an E.C. between about 2500 and 10,000 μS. An alternative definition is water with a TDS of between 10^3 and 10^4 mg.l^{-1}.

braided stream / channel An anastomosing stream (multiple channels that diverge and reunite around innumerable sand bars, with very complex 'braided' geometry) characteristic of very heavy sediment loads debouching from piedmont areas onto a plain. Such channels are shallow and unstable. The most famous are the Waimakarere and Rakaia rivers in South Island, New Zealand.

breakaway point = **breakway point**. The threshold velocity or flow rate at which a current meter or flowmeter begins to respond ($\neq$ 0 due to friction).

break-point chlorination A complex reaction to remove ammoniacal nitrogen by the addition of chlorine in excess of that used to break down combined chlorine species (especially chloramines). Wastewater typically requires a dose of about 8 mg.l^{-1} Cl° to yield a chlorine residual of about 1 mg.l^{-1}. Breakpoint chlorination is the most commonly used method of converting combined nitrogen to free nitrogen, thus denitrifying wastewater. *See* **Chlorine residual**.

breakthrough time Given a point source of groundwater pollution introduced at time t = 0, the elapsed time before the first appearance of the pollutant at some down-flow point is the 'breakthrough time'. Note that the apparent flow rate of the pollutant does not necessarily = the Darcian velocity.

breakthrough curve A graphical depiction of the concentration of a pollutant as it moves through an aquifer. It may take the form of concentration *vs.* distance, or concentration *vs.* time.
Examples of breakthrough curves:

$\dfrac{C_t}{C_o}$ strong retardation weak retardation

distance = v.t →

brick wall scenario A notional fixed limit of 100% development of all available water resources. In practice the 'brick wall' is never reached because cost, social, political and environmental constraints arise well before 100% water usage can be attained. Moreover, the 'brick wall' usually recedes as it is approached because economics make desalination progressively more attractive.

bridge The clogging of pores by small particles in an aquifer, filter pack or screen.

brine Water of salinity similar to or greater than sea water, such as in salt lakes, deep groundwaters, some mine waters, and ground-waters associated with oilfields.

broad-crested weir A masonry flow measuring structure in which the downstream weir section is that of a free-fall nappe. It is designed to work under supercritical / critical flow conditions.

Dictionary of Hydrology and Water Resources

Brunt-Väsälä Frequency This frequency, N, is a measure of the stability of a thermocline. A large value of N results in virtually no mixing of the epilimnion and hypolimnion (bar thermal and ionic diffusion). A small value of N indicates that even minor wind shear may be sufficient to overturn and mix the layers. $N = \sqrt{\dfrac{g}{\rho}\dfrac{d\rho}{dz}}$, a function of the vertical density gradient, mean density and g. It is normal to quote N^2.

BTEX = **Benzene, Toluene, Ethylbenzene and Xylene**; sometimes referred to as BTX. These are the important water-soluble compounds in petroleum. Traces of any or all of these are characteristic (even diagnostic) of groundwater pollution by oil.

bubble gauge
(1) A stage measuring device utilizing a controlled release of gas from a stream bed. The back pressure is proportional to the stage.
(2) A means of measuring the velocity profile of a stream or small river by the lateral displacement of gas bubbles as they rise from the stream bed. *See* **RAFT**.

bulk density The overall density of a substance, such as aquifer material, inclusive of saturated and unsaturated voids; the apparent density of wet soil or rock.

buoyancy pump = Air lift pump.

buried channel / valley A fluvial channel which has been filled in and subsequently obscured by sedimentation. Such channels are often of much higher hydraulic conductivity than the surrounding rock / sediment, and are therefore good targets for groundwater exploration. There are several geophysical techniques for locating them.

bushfire = wildfire In addition to the obvious hydrologic effects of vegetation loss, bushfires introduce oxidized organics into the near-surface soil horizons, which tends to make the affected soils hydrophobic (water repellent). A comparable condition also occurs in weathered or otherwise exposed peat.

buttress dam = 'hollow dam', or 'Ambursen dam'. A once common form of masonry dam in which a thin (relative to a gravity dam) concrete wall is supported on the down-stream side by a series of buttresses. It requires up to about 60% less concrete, but a lot more steel reinforcement. Buttress dams are best suited to wide valleys where solid rock is a limiting factor.

bypass flow Flow through a flood relief channel.

caatinga A geomorphic term from the Amazon basin: It is a distinctive type of tropical rain forest consisting of nutrient-poor quartz sandy soils and their associated vegetation. River water draining from caatingas is typified by a high TOC from humic and other organic acids.

cable rig = Percussion rig. A method of water well drilling with a reciprocating rock chisel. The method is slow, and normally limited to <100 metres' depth. However, it is simple, cheap (relative to rotary drilling), and yields good geological control.

caisson An open chamber, often cylindrical, sunk in water and pumped dry to gain access to a stream bed, portion of estuary sediments, for example.

cadmium 'Cd'. A toxic heavy metal geochemically similar to zinc, but about 200 times less abundant. Until the late 1960s human exposure to cadmium was widespread from air, water and food ingestion, especially in industrial areas. In Japan it was realized that itai-itai disease (=Ouch! Ouch!, referring to the pain of joint disease) was caused by eating rice grown from irrigation water with high concentrations of Cd. In Sri Lanka there is continuing suspicion that Cd, concentrated from phosphatic fertilizers, may be one of the co-factors leading to acute renal failure. In most countries environmental remediation over the last few decades has resulted in about a ten-fold decrease of Cd concentrations in affected rivers. Cd solubility, usually as Cd^{2+}, is strongly pH controlled.

calcareous = **calciferous**; literally containing calcium. In practice, meaning calcium carbonate, for example, calcareous sandstone (= calcarenite).

calcium Ca^{2+} is a major cation found in almost all natural waters, in concentrations typically varying from 5 to 200 mg.l^{-1} (or much higher in some gypsum-rich environments). Limestone, gypsum and Ca-silicate dissolution are the main sources. Calcium, in combination with the bicarbonate ion, is the source of 'sweetness' in hard-water areas. In general, high calcium concentrations in water are beneficial, with the rare exceptions of some renal diseases.

calcrete = **caliche**. A natural carbonate cemented sub-horizontal horizon in the weathering profile of a soil or regolith. Calcretes are found world-wide, but are most common in the surface or near-surface vadose zone of arid and semi-arid limestone areas. If cementing material is gypsum (calcium sulphate dihydrate), rather than calcium carbonate, then the resulting rock is more correctly termed 'gypcrust' or 'gypcrete', although the term calcrete is often inappropriately used.

calibration curve = **rating curve**.

calibration tank = **towing tank**. A long straight tank with a moving towing trolley, to which a current meter is attached. A moving meter in static water is the only way that such a meter can be accurately calibrated. There are so few calibration tanks in operation that impellor-type current meters everywhere tend to be insufficiently recalibrated (and therefore inaccurate), especially in the developing world. This would be unimportant if everyone switched to using solid-state current meters.

calibre Sometimes 'caliber'; the inside diameter of casing, pipe or piezometer.

caliche The general meaning is 'calcrete' *(see above)*, but in Chile, it has the specialized meaning of sodium nitrate. In view of the inconsistency in the literature, it is best to use the term 'calcrete' for the more general connotation.

caliper A variety of borehole log which indicates the variation of diameter with depth. Large diameters generally correlate well with uncemented or poorly-cemented sediments, or with fractures. Small diameters are caused by hard beds or swelling clays. The caliper probe has between one and three sprung arms, pads or bows.

CAM	Crassulacean acid metabolism, or 'dark CO_2 fixation'. An unusual photosynthetic process in which plants transpire during the night using the 'C4-like' mechanism of malic acid ⇔ carbohydrate alternation. This minimizes moisture loss during the hottest parts of the day, so CAM species are well adapted to arid or semi-arid conditions. Examples of CAM plants include most cacti and succulents, gentians, saxiphrages, euphorbias, orchids, aloes, and yuccas.
canal section	A full description of a transverse vertical section of a canal, including bank type and slope(s), water levels, lining material, and underlying geology.
candled ice	River or lake ice with a vertical columnar structure.
canopy	The main foliar layer in a forest: Typically at 30-40 metres in a tropical rain forest, lower in a temperate forest, and may be near ground level, as in the Australian pygmy mallee and some alpine environments.
canopy interception loss	That fraction of the intercepted precipitation in the canopy which is lost to evaporation, (excluding drip and stemflow). This is usually a major factor in the water balance.
canopy resistance	Apparent resistance to the vertical flux of water vapour caused by 'leaf damping' in a forested area.
canopy storage	The amount of intercepted precipitation that can be temporarily stored on or in the foliage. For a mature rain forest emergent, this could be >100 litres per tree. Other quoted examples are 11 litres for a eucalypt, and 54 litres for a pine, but values vary greatly according to growth-stage, exposure, wind etc.

capillarity The effects of surface tension in drawing water up into narrow pores (against gravity). An air-water interface is normally assumed, but different capillary effects occur with other fluids, such as gas-oil.

capacitance probe Or capacitance monitor. Down-hole water-monitoring devices come in two varieties, for soil moisture and for water table monitoring. Soil capacitance probes can be up to about 120 cm long. They are placed in a pre-set monitoring tube in the soil, and measure the dielectric permittivity of the soil. Since the dielectric of water is >> soil, the probe effectively measures the water content of the soil. Soil properties vary, so the capacitance must be calibrated for any given soil. Capacitance probes are widely used in irrigation management.
Much longer capacitance devices can also be used to log short- or long-term changes in the water table.

capillary conductivity The 'constant' of the Darcy flow equation. Under saturated conditions, it really is a constant. Under *unsaturated* conditions, the capillary conductivity is orders of magnitude less, and inversely proportional to the soil moisture content.

capillary fringe = **capillary zone**. In a water table aquifer, the capillary fringe is that water held above the atmospheric pressure surface by capillarity, but not including the specific retention. The thickness of the capillary fringe varies from about a mm to several tens of cms.

capillary lift The height to which water will rise in a sediment or capillary tube, relative to a free water surface. For a glass tube in pure water at 20°C, the lift = $1.5/r$ mm where r is the tube inside radius.

capillary water Water held by surface tension.

capillary zone *See* **capillary fringe**.

capture *See* **river capture**.

carbonate hardness = Temporary hardness, calculated as: $50.(HCO_3^-$ meq $+ CO_3^{2-}$ meq).

carbon balance ≈ Carbon cycle. The very complex but roughly cyclical series of processes involving atmospheric CO_2, organic and inorganic carbon fluxes in the environment.

carbonic acid H_2CO_3. Rainfall dissolves atmospheric, and subsequently soil, CO_2, to form a weak solution of chemically aggressive carbonic acid, which dissociates to H^+ and HCO^{3-}.

carbon filtration Water filtration through GAC or PAC to remove a wide range of gases and organic impurities.

carbon - 13 ^{13}C; a stable heavy isotope of carbon which is sometimes helpful in interpreting ^{14}C data. For example, the $^{12}C/^{13}C$ fractionation of plants varies according to the photosynthetic pathway taken.

carbon - 14 A widely used method of dating groundwater (and other matter) by comparing the current concentration of the radioisotope ^{14}C with its probable concentration at the time of chemical evolution (or life cycle, precipitation, etc), and calculating the time taken to decay from the higher concentration to the lesser concentration. By very precise accelerator mass spectrometry dead carbon measurements can *appear* to be as old as ~80,000 years. In reality any radiocarbon date older than about 55,000 is meaningless because the measurement becomes less than the error bound intrinsic to the method.

carboxylic acid The most important functional group in natural dissolved organic carbon: -COOH. It is a very common component in humic and fulvic substances, and has an important role in complexing metals and rendering organic ions soluble.

carcinogen A cancer-promoting or -causing chemical. There is some concern about the risks of carcinogens in water, even though they are usually at ultra-trace concentrations (ie ppb or less). They include trihalomethanes, nitrosamines, most pesticides, and a wide range of PAH's. However, the risk should be kept in perspective: carcinogens ingested via food or smoke inhalation present much greater risks than carcinogens in water supplies.

carryover storage = **over-year storage**.

cascade A series of artificial or natural rapids.

cascade of errors A sequence of operations in which a small error in the first operation grows with succeeding operations, often leading to a catastrophically wrong conclusion. For example, a small error in estimating evaporative loss or mean aquifer porosity may give rise to a massive error in estimation of the ultimate available water resource. Computer codes and models of chaotic systems are prone to cascades of errors.

casing The tubular lining of a borehole, usually low-carbon steel but can be stainless steel, GI, ABS, PVC or fibreglass. Additionally, several 'special purpose' casings are available. Dug wells may be built with a stone or porous concrete casing.

catalysis The speeding-up of a chemical reaction by adding a substance (catalyst) which depresses the free energy

Dictionary of Hydrology and Water Resources

of activation. The catalyst does not itself change composition.

catalytic model A fine-sounding liberal, trendy, social-science concept of leadership in which water resources management is withdrawn from the domain of programme-driven experts, and given to water users and related stakeholders to address public issues collaboratively, (*cf.* **IWRM**). This may, perhaps, work in well-educated, well-resourced, highly motivated, corruption-free, democratically mature and socially responsible communities. Elsewhere, as in most developed and all developing countries, the catalytic model of water leadership/management nearly always fails dismally.

catch basin A sump, often built into an urban roadside gutter, to collect leaves, miscellaneous plastic, dead animals, *etc*. A high off-take in the sump conveys runoff to a stormwater drain.

catchment Synonym for "basin"; hence catchment area = basin area.

catchment yield The surface runoff at an exit point from a catchment. Usually expressed as the ratio of annual runoff to catchment area, in units of mm (or occasionally as a fraction of the mean annual precipitation).

catenary The curve, of the form $y = a + b.\tanh x$, formed by a suspended cableway, which may be taken into account when estimating an air line distance in stream gauging.

cathodic protection Protection of steel casing, screen, pipes *etc*. from corrosion, especially in warm, wet and/or saline conditions. This is achieved by a low voltage current

cation Any positively charged ion. Most metallic ions in solution are cations although complexed metal anions also occur, generally at low concentrations.

cat's paw A fleeting disturbance caused by a gust of wind on an otherwise calm water surface.

Cauchy boundary condition A boundary condition in which, if the head is known, the flux across the boundary can be calculated, or vice versa. One of several types of boundary condition used in defining ground-water models.

Cauchy number $= \text{Mach number} = \dfrac{v}{\sqrt{E/\rho}}$ where E is the bulk modulus of elasticity, ρ is the density, and v is the water velocity. This parameter relates inertia to elastic compressive forces. A constant Cauchy number is a requirement for similitude in water-hammer models and a few other specialized applications.

Caulweld auger A truck-mounted barrel auger for drilling wide-bore wells in soft sediments. The maximum depth of such a well is about 35 m.

caving (1) Partial to total collapse of the sides of a borehole.
(2) Stream bank collapse due to erosion by fast flowing water, especially on the outer bank of a sharp bend.

cavitation Reduction of hydrostatic pressure to below water vapour pressure results in bubbles of water vapour which subsequently implode on the moving surfaces of pumps, turbines, and other rotodynamic machines.

connected in such a way that electrons flow from the electrolyte to the metal. *See also* **sacrificial anode.**

Dictionary of Hydrology and Water Resources

This process of cavitation can cause severe erosion and mechanical problems. Cavitation can also occur naturally in fast flowing groundwaters under karstic conditions, and can be a significant erosive process.

cavitation index An index used to determine the maximum elevation above tailwater at which a turbine can be set without risk of cavitation damage. The cavitation index, $\sigma = \dfrac{\Delta H}{h}$ where ΔH is the head relative to atmosphere - velocity head - the potential head at the turbine relative to the tailwater, and h is the upstream effective head on the turbines.

cdf Cumulative density function. The integral of a probability density function. The cdf is a step in the derivation of any probability distribution.

CDTA A highly specific complexing agent which binds with Al and Fe, thereby reducing interference in the ISE analysis of fluoride.

CEC (Cation exchange capacity). A measure of any mineral's capacity to exchange cations with those in solution. Measured in meq. per 100 g. Clays, some organic molecules, and a few other minerals have a sufficiently large CAC to profoundly affect the chemistry of any groundwater in contact with the mineral. CECs are pH dependent. Typical values, in meq.100 g^{-1}, are: humic acids 400, smectite clays 60 to 150, illite clay 10 to 40, kaolinite clay 3 to 50, fine-grained mixed stream sediments 5 to 50.

CECs 'Compounds of emerging concern', generally trace organic compounds with 'gender-bending' properties (endocrine disruptors), which wreck fish and amphibian ecology, and which are suspected of being human health hazards. Other CECs which find their way into the aquatic environment include some carcinogens, veterinary products, and a broad spectrum of antibiotics and other medicines. Unfortunately, many CECs are hydrophilic, and are not susceptible to biodegradation.

celerity   The speed of propagation of a wave, upstream or down-stream.

cenote   A karstic collapse feature associated with a high water table and resulting in a small subcircular lake. The diameter: b depth ratio is usually ≤1.

centralisers   Sprung steel plates welded or clamped to borehole casing to keep it straight and vertical during emplacement.

centre of pressure   The point at which a distributed force can be considered to act as a single force. Also, the point in a dam profile at which the mean hydrostatic pressure can be considered to act.

centrifugal pump   A variety of long narrow-diameter pump used in boreholes. The water is forced up the attached rising main by centrifugal force, imparted by a high speed axial impellor.

chalybeate spring   *(Obsolete term)* Unusually iron-rich emergent groundwaters, such as those caused by pyrite oxidation within the aquifer: $FeS_2 + 4O_2 \rightarrow Fe^{2+} + SO_4^{2-}$. The iron typically oxidizes further within the surface discharge zone to precipitate a characteristic brown precipitate of Fe-oxy-hydroxides.

chamber   A huge karstic cavern.

Dictionary of Hydrology and Water Resources

channel accretion The shallowing of a channel by sedimentation. A serious problem in rivers such as the Yellow River (China), and the Mississippi (USA).

channel capacity = **bankfull capacity**. The cross-section of a channel, in m^2. Though normally obvious, there are some compound channels where it is a moot point as to what constitutes the channel capacity.

channel control = **partial control**. The condition in which the stage is not proportional to discharge. The hydraulic gradient is influenced by upstream / downstream conditions, *cf.* 'section control'.

channel frequency The number of stream segments per unit area.

channel improvement Various works designed to even or stabilize flow, or to avoid erosion and sedimentation. Examples include stone revetment, steel piling along a bank, or straightening a reach.

channel line The plan locus of maximum velocity of a stream or river. For most purposes, the thalweg and channel line may be regarded as the same.

channel loss Water loss through seepage. The main source of water loss in many irrigation systems.

channel maintenance A little-used parameter defined as land area per unit channel length.

channel precipitation That precipitation which falls directly into surface water, and which therefore responds as instant runoff. It is seldom more than a percent or two of total precipitation, except in lakes, seasonal floods, swamps, *etc*.

channel roughness Nowadays, the Manning roughness coefficient. Previously, the Chezy roughness coefficient.

channel span A span across the deepest part of a river; that part of a bridge most crucial to river traffic.

channel storage A parameter that must be known for each reach of a river in order to calculate the shape of a flood hydrograph as it progresses downstream (the procedure of flood routing).

check irrigation Irrigation from the top of a field, as controlled by a series of 'checks' or ridges.

chelate A complexed molecule where a 'ring' or quasi-cyclic structure is formed between a metal ion and two or more electron donors from an organic ligand. Contrary to some texts, not all organo-metallic complexes are chelates.

chemical oxygen demand COD and BOD are the two most commonly used measures of water pollution by reduced species, mainly organic.

chemocline The transition zone from lower salinity to higher salinity layers in a lake; from an upper low-density layer to a deeper high-salinity layer. This chemical stratification is analogous to the thermocline in a thermally stratified lake, except that there may be more than one chemocline at varying depths in the same lake.

chemotaxis The 'swimming' movement of micro-organisms towards higher concentrations of oxygen or nutrients.

chemotropism The growth of microorganisms towards higher concentrations of oxygen or nutrients.

chevron drain = Herringbone tile drain.

Chezy equation For uniform flow: $Q = A\bar{v} = AC\sqrt{R.S}$, where Q is the discharge, A is the cross sectional area of flow, C is the Chezy roughness coefficient, R is the hydraulic mean radius, and S is the channel slope. This is an empirical equation for estimating the mean velocity of open channel flow from the channel morphology. Some authorities still prefer the Chezy equation despite its inferiority to the Manning equation.

Chezy coefficient (C) An empirical 'roughness' or 'friction' factor, required for the Chezy equation. It is more dimensionally consistent than the Manning coefficient (n), but is less accurate and too variable for good field practice. Note that: $nC = R^{\frac{1}{6}}$

chloramination Disinfection of a water supply by the addition of both chlorine and ammonia. Chloramination is less effective than normal chlorination, but the effects last much longer. Thus chloramin-ation is well suited to use in reticulation systems with a long residence time, such as long pipelines.

chlorination The practice of treating water supplies with low doses of free chlorine. This is a strong oxidizing agent and bactericide.

chlorine residual The more or less stable concentration of $Cl°$ in water after much of the original chlorine dose has been neutralized by ammonia, organics and bacteria. For effective disinfection of water, enough chlorine must be added to leave a residual of at least 0.2 mg.l^{-1} 10 minutes after chlorination. In practice, the chlorine residual typically varies from about 0.2 to 1.0 mg.l^{-1}. Initial disinfection, such as treating a bore, pipe or tank prior to commissioning, requires a higher residual of about 10 mg.l⁻¹.

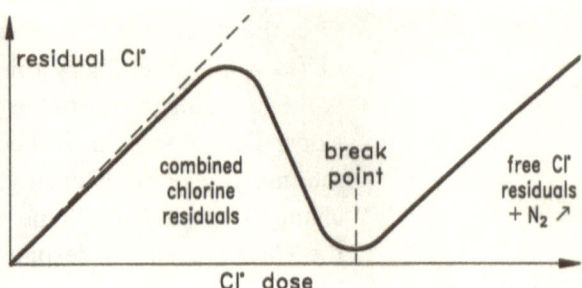

| chott | A North African seasonal lake, usually saline. The chott bed has a high albedo, and commonly consists of evaporitic minerals such as halite, gypsum or anhydrite. Examples include Chott el Djerid in Tunisia, and Chott Melrhia in Algeria. Some chotts have areas of thousands of square kilometres. |

chronic renal failure A painful and life threatening health-hazard assumed to be associated with polluted drinking water in parts of central, north and eastern Sri Lanka (and, rarely, elsewhere). Once thought to be caused by cadmium from phosphatic fertilizers, which is concentrated in cascade irrigation systems. More recent research suggests that CRF is not so simple, and probably has a multi-factorial aetiology.

Cipoletti weir A trapezoidally-shaped sharp-crested weir, designed to minimize the effects of end contraction.

cistern A small water storage reservoir recessed into the ground, and often covered to minimize pollution and evaporation. In ancient times, primitive cisterns were widely used for domestic storage and dryland agriculture. They are still used in some parts of the world.

Dictionary of Hydrology and Water Resources

class A pan	A standardized evaporation pan, 1.22 m in diameter and 0.25 m deep used to estimate rates of evaporation from a large open body of water (actual or hypothetical). Since evaporation is different between macro- and micro-climatic environments, the method is imprecise. A real lake evaporation rate = the 'class A pan' evaporation x a pan coefficient (typically between 0.60 and 0.75).
clasticity index	The maximum grain size of a sediment.
clastic rock	Literally, a rock consisting of 'broken fragments'. Any rock consisting of a conglomeration of particles, such as sand-stone, breccia, conglomerate and tillite.
clay	*Engineer's viewpoint:* Any fine-grained sediment with both cohesion and internal friction, *or* any rock particle with a mean grain size of < 2 microns. *Mineralogist's viewpoint:* Any hydrous silicate with a layered molecular structure and either tetrahedrally or octahedrally co-ordinated cation layers. From a hydrogeological perspective, the less clay in the aquifer, the better. Clay clogs pores and well-screens, reduces permeability and storage, and, with progressive loss of aquifer pressure, carries the risk of hydrocompaction.
clay line	*See* **shale line**.
cleat	A set (or sets) of fractures perpendicular to the bedding plane of a coal seam.
clogging	Filtration of small particles through a membrane clogs by several different mechanisms: initially, blocking filtration occurs, followed by cake- or gel-filtration without compaction, and finally gel-filtration with compaction.

closed drainage = **internal drainage** = any endorheic system.

cloud base The height of the base of a cloud layer, which may be expressed in metres or equivalent pressure in hectopascals. This may be measured in three ways: (a) by measuring the reflectance time of a laser pulse, (b) By measuring the light intensity of a normal light beam by means of a ceilometer, or (c) by measuring the temperature of the cloud base with an infra-red thermometer. This is then compared with the ground- level shade temperature. The difference in temperature uses the average lapse rate of ~6.5°C.km^{-1}, to estimate the height. Naturally, the latter is only applicable if there is no significant surface-base inversion. Manual errors in estimating the cloud base can easily be ±20%, even by experienced observers.

cloudburst Strictly, any rainfall intensity ≥100 mm. hr^{1}. Such a rate is seldom attainable for more than a few minutes. More generally, any extremely intense rainfall.

cloud forest = **fog forest**. Wet forest in perennial or near-permanent cloud drape. This typically occurs in the mountainous tropics such as parts of the Peruvian Andes, Rwenzoris of central Africa, Papua New Guinea, etc. Such areas tend to have extraordinary biodiversity, and act like 'sponges', slowly releasing important water supplies to lower areas downstream. Global warming is rapidly forcing cloud forest ecosystems to higher elevations, thereby reducing areas of unique habitat, and desiccating the lower base of cloud forests. In places like Indonesia and the Philippines, the decline of cloud forest is

cloud forest cont. compounded by logging and general land clearance. Only about 1% of the global woodland consists of

Dictionary of Hydrology and Water Resources

cloud seeding   = **Rainmaking**. Several techniques whereby finely dispersed AgI or 'dry ice' is injected into cold cloud tops (by plane or rocket) to initiate raindrop nucleation, and hence cause rain. There is no doubt that, under ideal conditions, it is possible to induce *local* rainfall. However, there is a great deal of doubt as to whether this is cost effective, or whether there is any nett gain in *regional* rainfall.

cloud forest, and the future prognosis for this habitat is bleak.

CLSA   Closed loop stripping analysis is a means of super-concentrating volatile trace organic pollutants, in preparation for analysis. This is done by refluxing air through a large water sample (using a sparging candle or bubbler), and passing the air through a tube of GAC to 'capture' the pollutant. It is then leached from the carbon with a few µl of methylene chloride.

coagulation   ≈ **flocculation**. The clumping together of suspended or colloidal matter as flocks, resulting in their precipitation. Coagulation may be induced biologically, or by the addition of a suitable chemical coagulant such as aluminium or iron salts, possibly with organic co-polymers.

coalescence   *See* **collision.**

cobble   Loosely, any cricket-ball sized stone. Technically, fluvial rounded sediment with a diameter in the range 64 to 256 mm.

Co-kriging   Kriging applied to more than one inter-related and spatially correlated variable.

coefficient of approach A 'fiddle factor' applied to flume and weir discharge equations to correct for errors caused by the velocity of flow just upstream of the structure. (The equations assume zero 'approach velocity').

coefficient of consolidation (C_v) A parameter used in the calculation of hydro-compaction, especially in clay-rich sediments. Usually C_v refers to the coefficient of *vertical* consolidation, = the vertical hydraulic diffusivity = the ratio of hydraulic conductivity to specific storage,

$$ie: C_v = \frac{K_y}{S_s'}$$

coefficient of contraction A correction factor applied to flume and weir discharge equations to correct for secondary flow in cases where the measuring structure is not built across the full width of the stream.

coefficient of discharge (C_d) Empirical coefficients applied to stage-discharge equations to compensate for friction and other systematic errors. C_d for a weir typically lies between about 0.58 and 0.75.

coefficient of evaporation *See* **pan coefficient**.

coefficient of fineness The ratio of suspended matter to turbidity, or the predominant particle size contributing to turbidity.

coefficient of groundwater discharge The ratio of groundwater discharge from a catchment (in equivalent mm) to total precipitation.

coefficient of leakage *See* **leakage coefficient**.

coefficient of regime For a river in any given year, the ratio of absolute maximum discharge to absolute minimum discharge.

coefficient of roughness ('n') Manning's roughness coefficient (see Manning's equation). Typical values are:

concrete or smooth earth channels	0.016 to 0.022
earth channels with weeds, bends etc.	0.025 to 0.033
sand and gravel channels (deep to shallow)	0.035 to 0.058
rock channels (smooth to rough)	0.025 to 0.050
streams cascading over boulders	0.030 to 0.070
overbank flow onto crops or fields	0.025 to 0.035
overbank flow onto long grass or woods	0.040 to 0.200

coefficient of skewness *See* **skewness**.

coefficient of uniformity (1) *Surface water:* A measure of the uniformity of water distribution in an array of irrigation sprinklers.
(2) *Groundwater:* An index of the spread of particle sizes in a sediment. From a sieve analysis curve, the CU is taken as: *40% retained size ÷ 90% retained size.* For a very uniformly sized sediment, the CU may be ≤ 2, the CU of a moderately uniform sediment may be ≤ 4. For a heterogeneous sediment the CU will be >5, but so variable as to be of little significance.

coefficient of variation $\frac{\sigma}{x}100\%$; σ is the standard deviation and $\bar{x}$ is the mean.

coefficient of viscosity = **dynamic viscosity,** = **absolute viscosity,** = **internal friction.** η or μ, in units of Poisseulle, PL, (= $kg \cdot m^{-1} \cdot s^{-1}$). A temperature-dependent intrinsic fluid property, which indicates the ease of shearing deformation or resistance to flow. Values of μ for pure water at different temperatures are:

1.80×10^{-3} PL at 0°C 1.01×10^{-3} PL at 20°C
1.34×10^{-3} PL at 10°C 0.80×10^{-3} PL at 30°C

coefficient of volume compressibility M_v. A parameter, usually measured in the laboratory upon fine-grained sediments, and used in estimates of hydrocompaction. It is determined as: $M_v = \dfrac{1}{(1+e_i)}\dfrac{(e_i - e_f)}{(P_f - P_i)}$ Where P is pressure head, e is the void ratio relative to 'solids =1' (so total volume = 1+e), and i and f refer to initial and final.

coffer dam Any temporary structure designed to hold back surface water from a work site.

cohesion Resistance of soil to deformation, caused by intergranular forces in clays and other fine-grained substances. It varies with moisture content.

cohesive soil A clay, soil or similar substance that deforms plastically under wet and dry conditions.

cold monomictic A class of lake where the surface temperature varies up to 4°C, in consequence of which it overturns (and mixes) once a year, during the Autumn.

coliform Several types of aqueous bacteria characteristic of, or even diagnostic of, fecal pollution from warm-blooded animals; whose presence indicates probable sewage pollution of a water supply. Coliforms consist of all aerobic and facultative anaerobic, non-spore forming, rod-shaped, gram-negative bacteria which ferment lactose with the release of gas within 48 hours of incubation at 35°C. A less rigorous, but perhaps more practical definition is: All bacteria that produce a purplish-green colony, with a metallic sheen, within 24 hours of incubation on an appropriate culture medium. Coliform bacteria are extremely common in the environment, but it is specifically 'repeated incidence of fecal coliforms' that is the first positive evidence of suspected water pollution. Some common species are: ***Escherichia coli*** *(q.v.) Enterobacter aerogenes*, and *Klebsiella pneumonia.*

coliform count The number of easily visible coliform colonies per ml of water, as counted from a culture plate after the appropriate incubation.

collar A concrete block which maintains water tightness around a metal-ceramic pipe junction. *See also* **drill collar**.

collection lysimeter A simple form of lysimeter consisting of a vessel dug into the unsaturated zone, in which any deep infiltration is collected in a sump, usually with an access tube to the surface.

collision After nucleation and prior to precipitation, raindrops, snow or hail particles grow by any of three collision mechanisms: coalescence, accretion or aggregation. *Coalescence* occurs in warm clouds (> 0°C) by the merging of fine water droplets, from as small as 0.2µm, to form drops of up to 7 mm. As the latter size is approached, they become aerodynamically unstable. Hence equilibrium is reached between coalescence and break-up. *Accretion* is the freezing of water onto ice to make snow, whilst *aggregation* is the clumping together of ice particles to form hail. The latter typically occurs in cloud tops at about -40°C.

colloid A substance 'not quite in solution', with a particle size in the approximate range of 5 to 200 nm. Colloids have properties intermediate between dissolved and suspended matter. Examples include inorganic polymers such as: $n.Al(OH)_3$, and a huge range of large organic molecules, such as humic and fulvic acids.

colluvium A predominantly coarse, but heterogeneous sediment comprising both gravitational and outwash material. Colluvium accumulates as scree and fan material in precipitous terrain. Colluvial permeabilities are usually high to enormous.

colour The *true* colour of water is blue, and is caused by the differential dispersion and absorption of white light. The *apparent* colour is due partly to reflection of suspended solids, and partly to the masking effects of dissolved pigments (mainly organic acids).

colour composite A reconstruction of a remotely-sensed image, usually from a satellite. The final colour picture is synthesized from three or more digital signals obtained from spectral filters in the original sensor. The composite may be true or false colour, and may be enhanced by computerized processing to produce spectral ratios or 'stretched contrast'. Colour composites can observe features such as changes in the depth of shallow lakes, and the extent of transpiration stress in crops, etc, which are not apparent from photographs.

colour unit (C.U. = Hazen unit). An obsolete measure of water colour (*see* 'absorbance') based upon an arbitrary standard potassium chloro-platinate solution. <10 C.U is scarcely discernible to the casual observer, whereas 50 C.U. is distinctly coloured. 'Tea-coloured' water from swamps attain about 200 C.U. A rough conversion to modern units is 14 C.U.s ≈ 1 Ab.m^{-1}.

combination well A dug well with horizontal collection adits at depth. Usually dug in areas of low permeability with sparse rock fractures.

combined chlorine The initial chlorine added to a water during disinfection tends to be 'mopped up' as combined chlorine, usually in the form of chloramines or chloro-organics. Increased dosage results in free chlorine. Combined chlorine may have some disinfectant action, but is not nearly as effective as free chlorine. *See also* **break-point**.

combined sewer A sewer receiving both sewage and stormwater runoff.

combined water Water which is chemically or electrostatically bound to water, and which can only be removed by heating.

common ion effect The decrease in solubility of a salt which results from one of the ions of that salt being already present in the water.

compensation flow Maintenance of a minimum level of streamflow downstream of a river control structure. Usually a legal requirement for down-stream water users.

competence The ability of a river to erode sediment from an upstream reach, transport it, and deposit at least the coarser fraction of sediment, further downstream.

complex ion A solute species formed by the pairing of hydrated cations and anions, for example: $Na^+ + SO_4^{2-} \rightarrow NaSO_4^-$ (aq). The effect of complexation is to lower the ionic strength of a solution. This becomes increasingly prevalent in more concentrated solutions.

compound aquifer = Composite aquifer. A layered alternation of aquifers and semi-confining layers in hydraulic continuity, (as opposed to a multiple aquifer in which the transmissive layers are effectively separated by aquicludes).

compound fan A series of coalescing alluvial fans along a fault or piedmont. Coarse-grained compound fans generally afford good ground-water storage capacity.

compound gauge A multi-compartmented ground-level raingauge in which the splash-in to the central compartment is assumed to equal the splash-in to the outer compartments, thus cancelling out the 'splash loss'. It is probably the most accurate method of rainfall measurement, but is impractical for general use.

compound hydrograph A multi-peaked hydrograph which can form either by the addition of tributary flood waves, or by complex patterns of precipitation across a catchment (or both).

compound hydrograph diagram

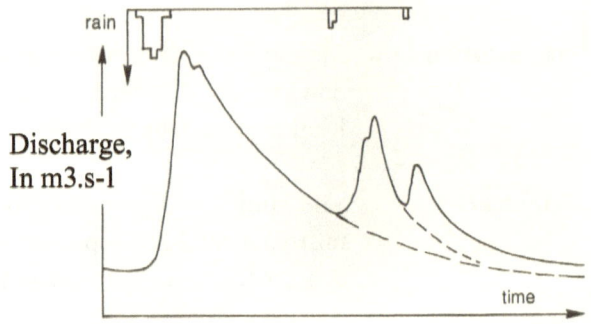

compound section A river section with markedly varying roughness coefficients.

compound weir Various designs of geometrically complex weir for measuring flows accurately under both high and low discharges.

compressibility of rock . . . or any other soil or sediment: $\beta a \approx 10^{-8}$ to 10^{-4} m^2.N^{-1}. As a confined aquifer is pumped (depressurized), loss of hydrostatic pressure *increases* the inter-granular pressure which causes hydrocompaction. It is this compressibility of the aquifer which accounts for most of the water yield from a confined aquifer.

compressibility of water $\beta w \approx 4.4$ to 4.8×10^{-10} m^2.N^{-1},(or Pa^{-1})—it varies a little with temperature. Water expansion by decompression is normally a small fraction of the yield from a confined aquifer. Water is about a hundred times as compressible as steel.

Compton scattering γ-ray scattering and energy degradation by interaction with atomic electrons. When γ-rays of between 0.1 and 1 MeV collide with rocks, Compton scattering is the dominant form of energy dissipation; a mechanism upon which γ- γ logging of boreholes is based.

computer model Nowadays ≈ digital model (as opposed to an analogue model). A mathematically conceptualized simplification of any static or dynamic system which may be solved numerically. Hundreds of models are available commercially to analyze such problems as: flood waves along a river, pollutant transport in surface and groundwaters, flow in complex reticulation systems, catchment responses to rainfall, and the piezometric configuration of aquifers. Correctly used, computer models are the best available tool for water resources management

comp. model cont'd. However such models are frequently misused, with output being accepted too literally. Computer models are good at helping the user to understand how a system works but, contrary to popular belief, they are not always accurate as predictive tools.

concave bank The bank of a curved reach of river, with the greater radius of curvature.

condensation nuclei Aerosol particles of clay, dust, soot, sea-salt, pollen *etc.* around which water droplets condense. Typical diameters are in the range 10^{-3} to 10 microns.

conductive analog(ue) An obsolete means of surface or groundwater flow analysis which utilizes the potential distribution on

conductivity *See* **electrical, hydraulic** or **thermal conductivities**.

cone of depression = **cone of influence**. The shape of the water table in dynamic response to point extraction, such as in a single well. In a normal aquifer test, radial symmetry in the cone of depression is assumed unless there is evidence of anisotropic hydraulic conductivity.

cone of impression The configuration of the piezometric surface around a point of recharge, such as an injection well. The inverse of a cone of depression. 'Virtual cones of impression' are used in calculations of drawdown near lateral aquifer boundaries.

confidence band = **confidence interval**, = **confidence belt**. The area between specified upper and lower confidence limits.

confidence limits On the graph of any sample of data, such as magnitude *vs.* probability, upper and lower confidence bounds can be drawn. These indicate the field in which the actual population is expected to lie, within a selected level of probability. In flood analysis, the confidence limits, *CL*, can sometimes be expressed in the form:

$$CL = Q \pm \frac{F.\delta.\sigma}{\sqrt{N}},$$ where Q is the discharge for a

given ARI, F is a frequency factor for the normal distribution for the selected probability level, δ is a parameter for determining the standard error of a given pdf, σ is the standard deviation of Q, and N is the number of events in the sample. Confidence limits must accompany any reporting of forward projections of flood, drought, etc. It is misleading and deplorably unscientific to skip this important step!

The opening fragment at the top of the page reads: electrically conductive paper. It is limited by the assumptions of homogeneity and (except in simple cases) isotropy, and is now only of historic interest.

schematic of confidence limits:

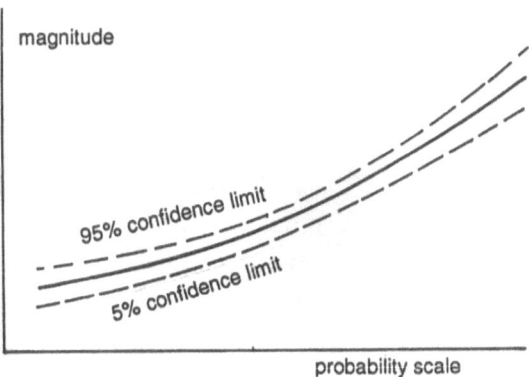

confined conditions ... In which the upper boundary of an aquifer consists of an impervious surface. Thus the upper surface of the aquifer is held at greater than atmospheric pressure, whilst the atmospheric pressure surface is within or above the upper confining layer.

confining layer = **confining bed**. The aquiclude which forms the upper bound of a confined aquifer.

confluence In the joining of two rivers, the combined flow reaches hydraulic uniformity almost immediately, whereas chemical and physical mixing may not occur until many km downstream. In the Amazon, some black and white river waters are still distinguishable some 40 km downstream of their confluence.

conjugate depths Water depths immediately upstream and downstream of an hydraulic jump.

conjugate shear Fracture sets oriented at 45° (in theory) to the principle stress axes in a tectonic regime. In fractured aquifers, one or other, or occasionally both conjugate shear sets may be important groundwater pathways.

conjunctive use / yield The combined use of all available water sources to optimize the available yield. This normally refers to surface water and normal groundwater storage, but may also include induced groundwater recharge, desalinated water, wastewater recycling, and 'water harvesting'. Careful manipulation of all sources during wet and dry conditions, taking account of groundwater recharge rates, and over-year reservoir storage, can yield more than the sum of individual yields of both resources.

connate water Groundwater that has long been isolated, by geological conditions, from normal meteoric circulation. Connate water is a small fraction of all groundwater, is commonly saline or otherwise mineralized, and may have a distinctive isotopic signature. It tends to occur at great depth where it is trapped by, or in, impermeable lithologies.

connate salt Residual salts (often marine brines), contained in pore waters since sedimentary deposition or diagenesis.

consequent drainage Drainage whose orientation is caused by flow down-dip along tilted strata.

conservation of mass One of the fundamental principles of science which expresses itself in many ways, such as:
(1) stream flow *cf.* the **continuity equation**.
(2) dilution gauging: Σ C.V.= K (sum of concentrations x volumes), "before and after", is constant, or

$$(3)\ S\frac{\partial \Psi}{\partial t} = K_x \frac{\partial^2 \Psi}{\partial x^2} + K_y \frac{\partial^2 \Psi}{\partial y^2} + K_z \frac{\partial^2 \Psi}{\partial z^2},$$

Change in storage due to a change in head with time = the three dimensional change in groundwater flux, assuming zero recharge.

Dictionary of Hydrology and Water Resources

conservative ion Any ion that does not readily react, adsorb or change species. Once in solution, it generally stays in solution. Chloride is the most conservative major ion.

consolidation Settlement of a soil in response to an applied load, resulting in increased bulk density, and decreased porosity.

constant injection Constant rate injection is one of two possible dilution gauging techniques in which $Q = q \left(\dfrac{c_t - c_d}{c_d - c_u} \right)$ where Q is the stream discharge, q is the rate of addition of tracer, C_u is the upstream concentration of tracer (if any), C_t is the concentration of the introduced tracer, and C_d is the concentration of tracer at some well-mixed point further downstream.

consumption The total global water consumption, including the agricultural, industrial, domestic and municipal sectors, has increased from about 1000 km^3.yr^1 in 1940 to 3900 km^3.yr^1 in 2013, and is increasing at a rate of 1.6% per year. Compare the current global growth rate of 1.1% per year.

consumptive use Any water use that removes water from a catchment, either by evapotranspiration, or by export to another catchment. Virtually all water uses must be divided into consumptive and non-consumptive components in order to formulate a water balance.

contact spring Any spring whose existence is structurally controlled by a basal impervious threshold.

continental divide The main watershed (English usage) which controls surface flow to one or other side of a continent.

The continental divide is obvious in mountainous divides, but ill-defined in places like western and central Australia.

contingent drought An unpredictable drought caused by weather variation rather than seasonality.

continuity equation For a uniform stream without tributaries or distributaries, $Q = A.v$ where Q is the discharge (mass per unit time), A is the cross sectional area, and v is the mean velocity. Alternatively, $A_1v_1 = A_2v_2$. In the case of influent or effluent streams, $Q = A.v + \delta S$ where δS is the change in groundwater storage.

contingency tables A statistical method for assessing the degree of association between two categorical ('either'/'or') variables. Contingency tables are occasionally of value in evaluating water development options when there are inter-related criteria.

continuous stream A stream that has continuity of surface flow throughout its reach, even though it may be ephemeral. Rivers in recession in arid / semi-arid regions are often discontinuous.

contour irrigation Irrigation onto contoured terraces (such as paddies), the object being to minimize soil loss and water runoff, whilst maximizing agricultural land use. Contour irrigation is one of many possible mechanisms for improving water-use efficiency.

contraction If a weir is constructed over a partial stream width, the flow over the weir is 'contracted' by a component of flow perpendicular to the channel line. This affects the coefficient of discharge.

control *See* **channel control**.

control well During an aquifer test the pumped well, and any observation wells, must be corrected for any natural fluctuations in SWL. This requires monitoring of a control well or bore in the same aquifer as the one under test, but beyond the radius of influence of the pumped well. A control well at about a km from the test is suitable, but the distance is usually determined by availability.

convection (In surface or groundwater). A reflux or continuous flowpath in a vertical plane, usually elliptical, and with or without inflow or outflow. There are several convective geometries, including single, multiple and toroidal cells. In *free* convection the flow is driven by density inversion (salinity or thermal). In *forced* convection the flow is 'stirred' by some form of pressure differential. Some groundwater systems are a mixture of free and forced convection.

convective precipitation = **air mass precipitation**. Precipitation from convective cells is relatively localized but sometimes very intense. Individual convective storms are only a few kilometres wide, but in aggregate, may account for a major fraction of regional rainfall.

conveyance loss General term for the water lost from irrigation channels and drainage ditches. Conveyance loss is mainly through seepage but evaporation and illegal extraction is sometimes included. There is no consistency of definition. Quoted units include $m^3.km^{-1}$, $percent.km^{-1}$, and $litres.s^{-1}m^{-2}$.

Cooper-Jacob equation An approximate transient solution for drawdown in a confined aquifer during an aquifer test:

$$s = \frac{2.3Q}{4\pi T} \log \frac{2.25Tt}{r^2 S}$$ where s is the drawdown, S is the storativity, Q is the constant discharge rate,

	T is the transmissivity, and t is the time since the start of pumping. This approximation is valid for $u < 0.01$ where $u = \dfrac{r^2 s}{4Tt}$. Therefore it should not be used for the first few minutes of the time-drawdown data.
copper	At normal concentrations copper is an essential element. Only rarely does natural dissolved copper occur at sufficiently high concentrations to be considered a major ion. The median value is about 0.01mg.l^{-1}, and the natural range is about 0.0005 to 1 mg.l-1. Technically, Cu is a toxic 'heavy metal', but the toxicity threshold is high; nausea and vomiting can occur at higher than 3 mg.l^{-1}, or slightly above the taste threshold of about 2.5 mg.l^{-1}. The latter is nearly an order of magnitude higher than the peak concentration used to treat algae in reservoirs. The acute lethal dose of copper for an adult is in the region of 30 grams. Cu contamination in drinking water is usually generated from copper plumbing rather than from the original water supply.
coprecipitation	Incorporation of a 'fugitive' salt in a precipitating mineral. For example the reaction: $BaCl_2 + SO_4^{2-} \rightarrow BaSO_4$ (insoluble) $+ 2Cl^-$ (used in the gravimetric determination of sulphate), may co-precipitate any carbonate in solution as $BaCO_3$ unless the reaction takes place at low pH.
coquina	Coquina beds are extremely porous and permeable calciferous sandstones composed of shell fragments, mostly poorly cemented.
core boring	A means of borehole drilling in which an annular bit recovers a relatively undisturbed core sample in a 'core barrel'.

corrasion Erosive action by transported stream sediment.

correlation coefficient The sample correlation coefficient, 'r' is a measure of how closely related one set of data is to another comparable set of data.

$$r = \frac{\sum_{1}^{n}(x-\bar{x}).(y-\bar{y})}{\sqrt{\sum_{1}^{n}(x-\bar{x})^2} . \sqrt{\sum_{1}^{n}(y-\bar{y})^2}}, \text{ or covariance}$$

(x, y) ÷ the square root of covariance (x). covariance (y). A perfect relationship results in $r = 1$. A perfect inverse relationship gives $r = -1$; no relationship at all gives an r of 0, or near- zero.

corrosion In the hydrological context, chemical erosion of rocks resulting in increased porosity, and the development of karst, or the chemical attack of groundwater upon metal casing.

cost-benefit analysis A comparison of present worth costs and benefits to evaluate the viability of a project (such as a rural water supply scheme). It is essential for assessing the economic worth, but a poor indicator of net human worth.

cover crop The primary purposes of cover crops are to introduce nitrogen into the soil and to enhance weed control. In addition cover crops can reduce runoff, and hence soil erosion, and reduce drying of the soil. 'Ploughing-in' the cover crop can enhance soil moisture, but care is required over planting time in relation to precipitation in order to maintain a nett positive soil moisture balance.

cover drilling The practice of drilling a small diameter hole ahead of adit drives to give advance warning of water-bearing zones.

creep groundwater flow around or under a structure.

crest (1) The spill threshold of a dam or weir.
(2) The maximum stage of a wave or flood.

criterion *. . . of water quality . . .* The maximum safe concentration of a substance as deduced from the best available, though frequently inadequate, data. A criterion may be a function of bioassay data, such as: application factor x $_{48hr}LC_{50}$. Water quality criteria and standards are not synonymous.

critical concentration The threshold aqueous concentration, in excess of which either the soil or irrigated crop, is adversely affected. This refers to Boron, chloride, TDS *etc.* It is difficult to define precisely.

critical depth The depth of an open channel at which critical flow occurs.

critical discharge (q_c) The discharge per unit width of channel, at which incipient sediment mobilization and transport occurs. *See also* the **Schoklitsch equation** and **Shield's diagram**.

critical distance The horizontal distance from a watershed at which overland flow is strong enough to move soil particles.

critical flow (1) Occurs at the 'critical depth' when the flow velocity increases from subcritical to supercritical; Froud number =1. A state in which the stream flow is just fast enough to prevent the upstream propagation of small waves. This is a necessary condition to maintain a unique stage-discharge relationship, and hence a pre-requisite for ease of flow measurement.
(2) Velocity at which flow changes from laminar to turbulent.
(3) For a given inlet head, the maximum discharge of a conduit with a free-fall exit.

critical gradient The hydraulic gradient corresponding to incipient mass flow or piping in a soil.

critical head The maximum applied pressure head to which a fine-grained sediment has previously been subjected. At applied pressure heads below the critical head, elastic deformation will occur. At an applied pressure above the critical head, plastic deformation, with slow water yield, will occur.

critical flow velocity The river velocity, v_c, at which incipient bed-load transport occurs. There are two commonly used ways to estimate v_c;
(1) the Manning-Limerinos equation, and (2) the Darcy-Weisbach-Hey equation, respectively:

$$v_c = \frac{1}{n} R^{\frac{2}{3}} . S^{\frac{1}{2}} \text{ where } n = \frac{0.113^{\frac{1}{6}}}{1.16 + 2\log\frac{d}{D_{84}}}$$

$$\text{and } v_c = \sqrt{\frac{8gds}{f}} \text{ where } \frac{1}{\sqrt{f}} = 2.03\log\left(\frac{ad}{3.5D_{84}}\right).$$

d is the depth, D_{84} is the particle size of the 84th percentile (the minimum diameter equal to or exceeded by 84% of the streambed particles), and a is a constant assumed to be 11.26. There are several variants of these equations. For example,

$$\frac{1}{\sqrt{f}} = 1 + 2\log\frac{R}{D_{84}}, \text{ or } \frac{1}{\sqrt{f}} = \left(1 - \frac{K_s}{10R}\right).2\log\frac{12R}{K_s}$$

where R is the mean hydraulic radius and K_s is the proportion of flow depth occupied by flow resistance elements.

critical period ... *of a reservoir* ... The emptying time from full to dead storage level, without spilling.

critical point The critical point of water occurs at 374°C and 22 MPa. Water at the Earth's surface never reaches the critical point, but may do so in deep geothermal aquifers in volcanic areas. As the critical point is approached the gas and liquid properties converge, so the distinction between water and steam becomes meaningless; however much the pressure is increased it is impossible for condensation to occur.

critical rainfall duration Peak discharge for a catchment will vary in response to short-period high-intensity rainfall, through to long-period lower- intensity rainfall. The critical rainfall intensity is that which results in the maximum peak discharge.

critical slope The gradient of a stream under critical flow conditions.

critical velocity = Belanger's critical velocity. For a given stream section, the velocity which marks the onset of critical flow; when a gravity wave just fails to propagate upstream, and when the specific energy of the stream is at a minimum.

crocodiles Despite the jokes, crocodiles can be a significant threat to hydrologists undertaking river gauging and other hydrologic measurements. Salt water crocodiles are common in the saline reaches of rivers in the Solomon Islands, Papua New Guinea, and parts of Northern Australia. Nile crocodiles are a threat in many parts of East Africa. In murky waters crocodiles can, and do, lurk almost invisibly, so take local advice before venturing into rivers in such areas.

crop moisture index The 'CMI' is an American variant of the Palmer drought index geared more towards assessing short term drought severity.

crop water use The rate of water consumption by transpiration, for example: as $mm.yr^{-1}$ or $mm.day^{-1}$.

cross-contamination ... *of aquifers* ... Drilling may create hydraulic continuity between two or more aquifers, in which case the lower pressure aquifers may be chemically contaminated by the higher pressure aquifer.

cross drainage Conveyance of surface drainage over or under roads. This may cost about a quarter of a road's construction budget.

crossover (1) *Surface water:* The point of inflexion at an S bend in a river, or a short reach of river between opposing bends.
(2) *Groundwater:* Discrete groundwater channels crossing at different levels in a karstic system.

crown drip Precipitation delayed by vegetative interception: difficult to measure accurately.

cryoscopy The measurement of a solute concentration by observing the depression of its freezing temperature.

Cryptosporidium A group of chlorine-resistant parasitic protozoans, common in faecally contaminated waters, such as runoff from farmland with a high stock density. With a size of 3 to 6 microns it is easily filtered out of water supplies, but is a common cause of diarrhoea worldwide, especially in young children. Most commonly the symptoms are relatively mild but, in the case of immuno-suppressed individuals infection can be life-threatening. Most species are harmless. However *C. parvum* and *C. hominis* are

	more troublesome species, and require either UV inactivation or removal by filtration.
CTD	An aquatic field instrument for measuring depth *vs.* electrical conductance and temperature, commonly used to monitor stratification in reservoirs. Some versions can also measure pH and dissolved oxygen.
CTT	Cloud Top Temperatures. CTTs have a major influence upon whether a cloud precipitates, and are usually measured by infra-red satellite imagery. The temperature threshold at which a cloud actively precipitates varies with climate and synoptic conditions, but is most typically between -50 and -60°C at the top of convective cumulus.
culvert	Surface drainage channel beneath a road or other structure. There are numerous designs based upon rectangular and circular sections. The hydraulics are not trivial, and depend upon the roughness coefficient, length, slope, size, approach velocity, entrance geometry, submergence, and entrance / exit conditions. The total head differential between headwater and tailwater is the sum of the entrance loss, friction loss and velocity head loss.
cumec	A largely obsolete term for the more formal cubic metre per second. $1 \text{ m}^3.\text{s}^{-1} = 10^3 \text{ litres.s}^{-1}$.
cumulonimbus	Dense towering clouds which spread out or disperse at high altitudes. May produce showers, hail or snow.
cumulus	Dense well-defined towering clouds which often originate at low altitude but may ascend to medium or high altitudes. Typically associated with storm cells and locally heavy rainfall.
cupola	*See* **solution pocket**.

curb	= Surface casing.
current meter	Any device that measures point water velocity in a stream.
curtain drain	Also known as **interception drain**. A drainage system, generally orthogonal to surface and/or groundwater flow, to divert water from a downstream area. Perhaps counter-intuitively, a line sink with a collector drain parallel to the flow, may be a more efficient drainage system than a curtain drain.
cusec	An obsolete imperial unit of flow. 1 cusec = 1 ft^3.s^{-1}. The cusec was long ago replaced by the metric 'cumec' (q.v.) except, perversely, in the USA.
cutoff	(1) Either an oxbow lake, or the 'short-circuiting' channel that produced it. (2) A grout curtain, clay, plastic or other impervious membrane in a dam or groundwater environment, designed to stop seepage.
cyanobacteria	Popularly known as blue-green algae, the cyanobacteria emit oxygen during photosynthesis. Most species are filamentous, and have the ability to fix nitrogen. They thrive in a wide variety of environments including marine, freshwater (notably rice paddies), moist nutrient-poor soils, wet rocks and in numerous symbiotic relationships. Their nitrogen yield is naturally low in polar regions, but rises to >100 Kg.Km^{-1}.yr^{1} in some tropical environments. Hot springs are often colonized by thermophilic cyanobacteria such as *Synechoccus lividus* (46 to 74°C). In desert soils cyanobacteria can fix sufficient NO$_3$ to create a water quality problem when it is leached to the water table. Although cyanobacteria are harmless at low concentrations their proliferation into blooms in freshwaters can pose a significant

health threat to humans and livestock, with potentially lethal effects from neurotoxins and hepato-toxins. In Asia, America and Europe more than 40% of lakes and reservoirs are now eutrophic. Such conditions are conducive to the rise of cyanobacterial blooms which disrupt or even destroy normal ecosystems. Species requiring control include *Aphanizomenon, Microsystis, Synechococcus, Nodularia* and *Anabaena*. The best form of control is to prevent waters becoming eutrophic in the first place. On the plus side, some edible cyanobacteria have demonstrable health benefits, such as anti-inflammatory effects and cytokine suppression.

cyclicity

There has been little success in identifying cyclic trends within hydrological data. Apart from the obvious seasonality, there are no consistently strong cyclic signals. There are only five credible cycles, and even these are statistically weak:
i 2.1 years: the quasi-biennial oscillation
ii 11 to 11.5 years: the solar (Hale) cycle
iii 18.6 years: the lunar nutational period (tidal)
iv 22 to 23 years: the double solar cycle
v 78 years: the Gleissberg solar cycle
I, ii and iv are observed in *some* long precipitation records. iii and iv are also known from the Nile runoff records, whilst v is known *only* from the Nile record (no other hydrological record is long enough). None of the above cycles are reliable for predictive purposes.

cyclic salt

Salt involved in meteoric circulation; precipitation of sea salt in rain, and its retention in solution until it drains back to the sea.

cyclone

A low pressure rotating storm system, associated with strong winds and heavy rain. The milder varieties occur in temperate latitudes, and are associated with frontal activity. Intense sub-tropical cyclones (hurricanes) are of entirely different nature and origin. (*NB:* 'cyclones' are 'anticyclones' in the southern hemisphere).

dambo
A geomorphic feature, especially associated with African plateaux, consisting of stream headwaters with no clearly defined channels. Instead, there is a pan-like swampy or marshland depression, with distinctive flora. Drainage from dambos has a distinct hydrograph shape.

dam failure
Of the numerous causes of dam failure the most important are: poor hydrologic analysis, insufficient attention to the foundation abutment and reservoir geology, poor maintenance, inadequate or erroneous design, poor construction (especially poor quality materials), extreme hydrologic anomalies, overspill & erosion of earth dams, earthquakes, rapid drawdown/ reservoir slope instability, internal erosion, and poor management.

dams
Worldwide there are more than 48000 dams over 15 meters high, with a gross surface water storage of about 6000 km^3, and an unknown but even larger associated storage of groundwater. Hydropower provides about 20% of the world's energy needs, whilst large dams feed irrigation water for about 17% of the global food production. Total expenditure upon large dams has been about $2 trillion, and is currently increasing at about $50 billion.year^1, despite both positive and negative environmental and human influences. Most of the best sites have already been utilized. Over 60% of all large river basins have some

Darcian velocity   degree of dam intervention. *See also* **rule curve**, and **displacement**.

Darcian velocity   Numerically equal to the *specific discharge* (groundwater flow per unit cross sectional area perpendicular to flow), = Q/A = -ki. This is the *apparent* velocity of groundwater flow, or the laminar flow rate that would occur if there were no granular obstruction in the flowpath. Naturally, since groundwater has to flow around particles, the actual distance of flow per unit time is greater than the direct line of flow, so the true water velocity is also greater than the Darcian velocity. *See* **effective velocity**.

Darcy flux   = Darcian velocity, = filtration velocity.

Darcy's law   The most basic groundwater flow equation in which discharge, $Q = -kiA$ where k is the hydraulic conductivity, i is the hydraulic gradient, and A is the cross-sectional area through which flow occurs. Since Q also = v.A, where v is the mean ("Darcian") fluid velocity, v = -ki. Darcy's law is valid only for steady (laminar) flow through a porous medium, and therefore fails at high fluid velocities. What is less well appreciated is that Darcy's law also fails at very low fluid velocities in fine-grained porous media.

Darcy Weisbach   The equation for friction head loss in a pipe: $h_f = f\left(\dfrac{L}{D}\right) \cdot \dfrac{v^2}{2g}$, the parameters being length, internal diameter, mean velocity and acceleration due to gravity. f is a function of Reynolds number, $\mathcal{R}$ and the 'relative roughness', $\dfrac{\varepsilon}{D}$, where ε is the roughness height of the pipewall. For laminar flow, $f = \dfrac{64}{\mathcal{R}}$, whilst for turbulent flow, f is a function of the relative roughness (independent of $\mathcal{R}$).

datum Or 'zero datum' is the reference level to which all survey levels are related, for example, in piezometry, or in river section surveys. It doesn't have to be with respect to sea level, but it does have to be consistent and fully referenced for follow-on surveys, which may be decades later. It's frustrating to find that a former datum either wasn't documented or that its bench-mark was demolished to make way for a new shopping centre.

Davies equation A modified form of the Debye-Hückel equation used to calculate activity coefficients in brines.

daya = **doline**; a silt-filled exokarstic feature, typically centred upon the intersection of limestone fractures. Although the limestone itself drains extremely rapidly, the surface silt-clay veneer may cause ephemeral ponding.

dead carbon Carbon devoid of the radioisotope ^{14}C. All but the most recent of carbonate rocks contain only dead carbon because they are very much older than the half life of ^{14}C, so any original ^{14}C has long since decayed to ^{14}N by β emission. ^{14}C dating of groundwater depends upon being able to deduce the proportions of dead and 'live' carbon at the time of hydrochemical evolution.

dead storage The volume of a reservoir below the offtake / exit level. Dead storage may be deliberately designed in a reservoir to allow for sedimentation. However, most sedimentation normally occurs in the active zone. *See* **active storage**.

dead time Any instrument that measures pulses (scintillations, waves, 'bucket tips', or electrical pulses) incurs dead time. This is a short interval during or just after a pulse, before the next pulse can be detected. When the count rate is large, the dead time may have a significant effect upon the instrument calibration.

débacle The spring break-up of river ice.

debris fan Also known as 'debris cone'. A fan-shaped depositional area of coarse- to fine-grained sediments, located at a break in slope, or in a piedmont area.

debris mark = **trash mark**. The 'high tide' mark left on a bank, overbank or tree after a flood peak has past, and from which the peak flow can be calculated. *See* **Mannings equation**.

Debye-Hückel equation $\log \gamma_i = \dfrac{-Az_i^2 \sqrt{I}}{1 + Ba_i \sqrt{I}}$, where γ_i is the activity coefficient of the 'ith' species in solution, a_i is an ionic size parameter (the Debye-Hückel radius), I is the ionic strength of the solution: $I = \dfrac{1}{2} \sum m_i z_i^2$, where m_i is the ionic concentration, Z_i is the ionic charge, and A and B are empirical constants. The activity coefficient is important in calculating solubilities of any mineral with respect to the water chemistry involved.

decay Most decay processes, such as radioactivity or some chemical reactions, conform or approximate to a decay of the form $C_t = C_0 . e^{-\lambda t}$, where C_t is

Dictionary of Hydrology and Water Resources

the concentration after time t, C_i is the initial concentration, and λ is the decay constant.

dedolomitisation The conversion of dolomite to limestone by diagenetic or meta-morphic exchange of Mg for Ca. The exact mechanism is unclear, but presumably requires an open through-flow of groundwater with a high Ca/Mg ratio.

deep karst A karst formation of which a significant fraction of the aquifer lies below the outflow point. Deep karst carries no implications concerning the thickness of the aquifer.

deep lead Fluvial channels deeply incised into bedrock during a period of river rejuvenation, such as during a period of tectonic uplift or marine regression. Deep leads often fill with more recent uncemented sediments to create good aquifers. Modern rivers may meander tens of kilometres away from deep leads.

deep percolation = **deep drainage**. Water that infiltrates through the soil or aquifer until it is too deep to be accessible by roots.

deep ripping Extra strong tines are used to deep-plough fields, especially for root crops, but also for other crops such as cotton. Advantages include the break-up of compacted soils, improved drainage, soil aeration, and increased field capacity. In flood-irrigated paddocks deep ripping combined with laser-levelling can greatly improve the water-use efficiency. Deep ripping is normally to a depth of 35 to 50 cm, but in some of the loess soils of Central Asia ripping extends as deep as a metre.

deep well 1) Any well designed for maximum drawdown, such as for foundations, quarrying or mine dewatering.

2) Any well whose dynamic water level is too low for a suction pump to be usable.

deflation   Wind erosion of fine-grained sediments. In the absence of vegetation, the ground level may be deflated until it reaches the water table, or until coarse-grained residua protect against further deflation.

deflocculant   = **dispersant**, = **dispersing agent**.

defluoridation   Several alternative processes used to remove fluoride from drinking water. Defluoridation agents include alum, activated alumina, and calcium salts. The latter exchanges to precipitate the highly insoluble CaF_2.

deforestation   This is arguably the single most important global influence upon hydrological processes. Fewer trees lead to less transpiration, less soil-moisture storage, more erosion, and faster and greater runoff. The effects of deforestation vary from the obvious to the unexpected, such as Australian bushfires releasing complex aromatic molecules (from gum trees) which permeate the topsoil, thereby rendering the soil hydrophobic, and resulting in more surface runoff.

degree days   Absolute degree days = the sum of mean daily temperature over a given duration: $\sum T_m \,°C$, and is used to determine the 'freezing index'. More generally it is the sum of mean daily temperature over a given duration, *above a threshold*, usually of 10°C. This gives an approximate measure of effective plant growth.

Dekhan   Central Asian term for a small plot used for growing household food + sale of any surplus, typically 0.35 to 0.50 ha. Such small crop areas are water inefficient.

delayed flow = baseflow.

delayed yield = Delayed water table response. An aquifer test in an unconfined aquifer produces a complex *log time vs. log drawdown* graph which is partly caused by slow dewatering - the delayed yield - from the unsaturated cone of depression. Semi-empirical type curves are required to analyse tests with delayed yield.

delf A drain on the landward side of a sea wall, levee or embankment.

deliquescence The property of a mineral to absorb atmospheric moisture. A few deliquescent evaporite minerals absorb so much water that they completely dissolve. Deliquescence tends to be very temperature and humidity sensitive. Many nitrates and chlor-nitrates are deliquescent, especially above an RH of ~ 60%.

delivery box A control and/or measurement installation within an irrigated channel system.

delta A roughly fan-shaped sedimentary deposit formed by a river debouching into a lake, estuary, *etc*. The sediments are generally well-sorted and porous, with variable and anisotropic permeability, that is with low horizontal permeability through foreset beds, but higher permeability laterally and along topset and bottomset beds. There is huge geomorphic variety of deltas, ranging from small steep coarse-grained fans, often incised, in piedmont zones, to massive fine-grained low-gradient fans with multiple active distributaries. The latter are associated with some of the world's largest rivers, such as the Nile, Mekong and Meghna-Brahmaputra; and

delta notation In isotopic work it is convenient to represent the heavy isotope in terms of enrichment or depletion relative to some universal standard. This is achieved by means of delta notation, where: $\delta\text{‰} = \left(\dfrac{R_{sample} - R_{std}}{R_{std}} \right)$, where R_{std} is the isotope ratio of the standard, eg $^{18}O/^{16}O$, or D/H. *See* 'SMOW' and 'PDB'.

demersal   Found at or near a stream bed.

DEM   Nowadays, a Digital Elevation Model is more or less obligatory in any Geographic Information System. It is generally a raster (array of pixels, each with an elevation) representation of a land surface. It can be interpolated from a contour map, or it can be estimated from satellite data, although in many parts of the world there may be difficulty in sourcing sufficiently precise and accurate primary data. A DEM is very useful in defining catchment and sub-catchment boundaries, and estimating catchment areas. But some cautions: In areas of low relief and low resolution of elevation, it is possible for major errors to occur in delimiting catchment boundaries. Always check the drainage distribution against the DEM catchment boundaries; there may be obvious errors that need to be resolved. Depending upon how the DEM was determined, it can be a 'digital terrain model' or a 'digital surface model'. A DTM approximates the actual land surface. A DSM is an approximate satellite view of whatever is reflected, including the canopy surface in forests, or the tops of buildings in urban areas. A DSM tends to be higher than a DTM. Finally, be aware that DEM elevations in mountainous areas can incur large errors.

dendroclimatology   Techniques of reconstructing past mean seasonal temperatures or precipitation records by using annual tree ring widths as proxy data. Variants: dendrohydrology and dendrochronology.

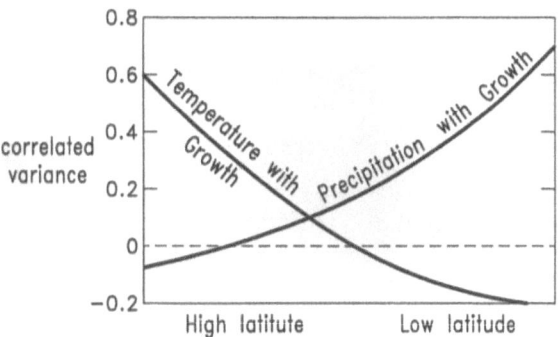

dendrometer An instrument for measuring the growth of trees by logging changes in circumference of the stem/trunk. Modern band dendrometers can detect changes in circumference of ± 0.005 to 0.01 mm, and hence record the effect of daily temperature changes as well as overall growth.

denitrification Bioremediation of nitrate contaminated water supplies can be achieved by lowering the DO to <0.3 mg.l^{-1} O$_2$, adding a suitable organic nutrient and reacting the NO$_3^-$ with non-pathogenic denitrifying bacteria. This is difficult to achieve on a large scale, so sometimes it is easier just to supply high-risk subjects, such as nursing mothers, with NO$_3$-free bottled water.

denitrifying bacteria Few heterotrophic bacterial species can metabolize the complete reduction of nitrate to molecular nitrogen. However, there are a great many species that, in combination, can achieve the task. These are about 150 known denitrifying bacteria, including *Paracoccus denitrificans*, *Pseudomonas (various species)*, *Thiobacillus denitrificans*, *Achromobacter*, and *Micrococcus denitrificans*.

density (of water) In SI units density, as Kg.m^{-3}, is 10^3 x the specific gravity. It is slightly temperature dependent, eg:

ρ (Kg. m^{-3}) at:	t=4°C	t=30°C
of pure water	1000	996
typical sediment-laden fresh water	1026	1020
mean sea water	1028	1022

density current Differential flow within a stratified water body.

density log = γ-γ log. A borehole log which measures backscattered γ rays from ^{60}C or ^{137}Cs sources. The detected signal is a function of bulk rock electron density, which is an analogue of absolute density and hence (indirectly) of porosity.

dentated baffle Tooth-like projections, usually in a spillway, to disperse kinetic energy. Occasionally also to aerate hypoxic or anoxic (hypolimnetic) waters from reservoirs. The baffles are incorporated into a 'dentated sill' (a concrete apron with protrusions).

depletion curve Different authors have used this as a synonym for the base-flow separation of a hydrograph, or the recession curve itself.

depression spring A water hole formed by a physiographic depression below the water table. Arguably not a true spring at all. An example is Alice 'Springs' - just water holes in the Todd River. *cf.* **oasis**.

depression storage Ephemeral water stored at the land surface. During rain, after the infiltration capacity has been reached, there is an interval prior to overland flow, during which depression storage is filled to capacity.

Depth-area-duration A family of D-A-D curves relates the depth of precipitation (mm), to rainstorm duration for varying catchment areas.

desalination = **desalting**. Energy intensive (and therefore very expensive) methods of removing salt and other solutes from sea water, brine or brackish water to provide fresh water. It is not economic to use desalinated water for crop irrigation. Desalination is an advanced technology, and employs such processes as flash distillation, reverse osmosis, and ion stripping.

desalinisation = **desalinization**. Reclamation of saline soils by such processes as tile drainage, flushing, the addition of appropriate soil additives, and more efficient irrigation. *Desalination* and *desalinisation* are often confused.

desert Regions in which potential evapotranspiration far exceeds the precipitation. The popular image of 'sand seas' ('Erg') is only one of many kinds of desert. Other landforms include flat gravel plains, barren mountains, and permanently frozen regions. Extreme desiccation results in a total absence of vegetation. Nevertheless, most deserts have a vegetative cover, albeit sparsely distributed, of highly specialized, drought-resistant species.

desertification The processes, both natural and human induced, by which land is gradually degraded to desert. These processes are complex, but may be summarized as either climatic change or excessive pressure from human activities such as nutrient exhaustion and overgrazing. It is an insidious, patchy process and, for most realistic and practical purposes, may as well be regarded as irreversible.

desiccation Drying out; particularly applying to falling water tables, decreasing annual precipitation, and desertification, due to both climatic change and human impact.

design discharge ≈ **design flood**. The maximum discharge which a hydraulic structure is designed to convey or contain. In dams, this is usually a flood peak of very large ARI (perhaps 50 to 100 years).

design storm The hyetograph used in deriving a time unit hydrograph (*see* **TUH**). In deterministically computing the storm discharge for a catchment, one has to make assumptions about the causal rainfall duration, total, and distribution within that duration. These assumptions constitute the design storm.

destratification Artificially destroying or impeding thermal stratification in a lake or reservoir in order to maintain oxic conditions, and hence good water quality. Techniques include stirring, whole column aeration and hypolimnetic aeration.

detection limit Strictly, the mean background signal 2 or 3 times the standard deviation of the background signal, (*cf.* 'lower limit of detection', 'limit of determination', and 'limit of quantitation'). This term has been badly abused by some scurrilous sales reps, who make outrageous claims for their analytical instruments.

detention period = **residence time**.

detention storage Transient storage, just sufficient to enhance groundwater recharge or to attenuate surface runoff.

detention structure A leaky dam or constricting structure built to achieve transient storage.

deterministic hydrology A holistic approach to understanding the hydrology of an environment by studying the individual processes that operate in that environment. The alternative is a stochastic approach, in which the output, such as floods, is considered to be generated from a 'black box' environment.

deuterium ^{2}H. 'Heavy hydrogen'. This 'heavy' but stable isotope occurs in average seawater at about 1 in 6420 atoms of hydrogen.

deuterium excess Rainfall whose oxygen isotopic concentration is that of SMOW (standard mean ocean water), will, on average, have a deuterium isotopic concentration in excess of SMOW. When expressed in δ notation, this excess is normally $\delta D = +10$, although there are regional variations, such as the Eastern Mediterranean where the excess is $\delta D = +22$.

developed length The mean length of flow, for example the centre line of a piped or surface channel route of flow.

developed water A legal term for surface water of artificial origin. For example, pumped groundwater, desalinated water, or water imported from another catchment. Some channel flows may be a blend of natural and developed water.

development ... *of a well* ... various processes of cleaning the fines and drilling fluid from a drilled well. This is undertaken after drilling, logging and placing the screen, but before any aquifer test is carried out.

dew point Temperature at which air of a given moisture content becomes saturated (pressure held constant).

dew point depression The temperature differential between ambience and dew point.

dewatering Pumping water away to depress the piezometric surface below an area of interest, such as for mining, quarrying, or foundation works.

diagenesis Post-depositional alteration of a rock. The many diagenetic processes include compaction, cementation, dolomitization, leaching, dissolution, recrystallization, bioturbation, and cryoturbation. The extent and nature of diagenesis has an important bearing upon the hydraulic properties of an aquifer.

diatoms A major sub-unit of the phytoplankton, responsible for about a 25% of the planet's oxygen production. These ubiquitous micro-organisms, typically 20 to 200 μm in diameter, are distinguished by having siliceous frustules (shells). You have almost certainly drunk billions of them without knowing it. They are normally harmless, perhaps even beneficial, although under conditions of extreme proliferation, a few rare species can be toxic.

differential permeability The effect of a semi-permeable membrane in permitting the flow of water molecules whilst barring the flow of larger molecules (most natural solutes), as in plant roots and reverse osmosis.

diffluence The process of glacial overflow across a ridge or col into another valley, or from a hanging valley into a deeper valley.

diffraction The bending of waves around an obstruction, such as a break-water or island.

diffusing well = recharge well, = injection well.

diffusion The molecular flow of water down a concentration gradient, causing spreading and dilution of a source of pollution. This occurs in both static and moving water, as opposed to dispersion, which is strictly a dynamic process. In practice, it may be difficult or impossible to distinguish between the effects of diffusion and slow dispersion, in which case the two processes are lumped together as 'hydrodynamic dispersion'.

diffusion equation An equation for fluid flow in a confined granular aquifer. If S_s is the specific storage, and K_n is the hydraulic conductivity in the n direction, then:
$$K_x \frac{\partial^2 \Psi}{\partial x^2} + K_y \frac{\partial^2 \Psi}{\partial y^2} + K_z \frac{\partial^2 \Psi}{\partial zx^2} = S_s \frac{\partial h}{\partial t}$$

If the aquifer is homogeneous then:
$$\frac{\partial^2 \Psi}{\partial x^2} + \frac{\partial^2 \Psi}{\partial y^2} + \frac{\partial^2 \Psi}{\partial zx^2} = \frac{S_s}{K} \frac{\partial h}{\partial t}$$
Note that the reciprocal constant, $\frac{K}{S_s}$, is the hydraulic diffusivity.

diffusivity *See* **hydraulic diffusivity.**

digestion Anaerobic biodegradation of sewage.

dike *See* **dyke.**

dilution gauging A stream gauging technique in which a tracer of known concen-tration is introduced into a stream, and its dilution is monitored at some well-mixed point further downstream. It can be applied in many situations, but is particularly effective in turbulent mountain streams of non-uniform section. *See* **gulp injection** and **constant (rate) injection.**

diluvium Obsolete term for coarse or partly-coarse glacial or fluvial sediment.

dimensional analysis A means of testing equations for consistency by reducing all the parameters to the three fundamental units of mass, length and time: [M], [L], and [T].

dimictic — Referring to a stratified lake which mixes twice a year; once in response to warming of the winter stratification, and again during Autumn cooling of the summer stratification.

dipole-dipole array — In field geophysics, a little-used configuration of 4 electrodes used for electrical resistivity soundings.

direct precipitation = **channel precipitation**.

direct pumping — The unsatisfactory practice of pumping directly to the end-user without intermediate gravity storage.

direct runoff = **quickflow**.

Dirichlet boundary — A fixed head boundary condition such as the coastal limit of an aquifer, or an aquifer in hydraulic continuity with a controlled river.

discharge = **flow rate**, usually in cumecs or $l.s^{-1}$.

discharge zone — The exit point of groundwater flow from an aquifer. It may be point discharge as in a spring, or diffuse discharge, as in a swamp or salt lake.

discretisation — The notional breakup of time and space continua into time steps and cell dimensions of uniform value for the purpose of numerical modeling.

disinfection — Elimination of pathogenic organisms from water, usually achieved by chlorination, UV or ozone exposure, *etc*. The final process of well, bore-hole or pipeline construction, prior to commissioning.

dispersant = **dispersive agent**. A Na^+ rich and / or acidic compound added to a clay suspension to prevent or reverse flocculation.

Dictionary of Hydrology and Water Resources

dispersion = **effective diffusion**, = the sum of molecular diffusion and inter-granular divergence. The dilution of a pollutant in flowing ground-water, caused by differential flowpaths and differing velocities in pores. *cf.* **hydrodynamic dispersion**, and **Peclet number**.

dispersive clay A clay in which molecular surface charges are balanced by H^+ or Na^+ from solution (as opposed to divalent cations). In this state, suspended clay will not flocculate, whilst deposited clay will exhibit a minimum permeability. For example, a dispersive clay would make a good lining for a land-fill site, to prevent the infiltration of pollutants.

displacement (1) 60±20 million people have been displaced worldwide by dam construction. About 1.5 million of these were displaced by the 'Three Gorges' dam alone. In most cases the displaced townspeople, villagers, and individuals have lost their homes and livelihoods, resulting in huge anti-dam sentiment and hence much more rigorous social planning in modern dam projects.
(2) The effective weight of a boat or ship, as measured by the equivalent volume of water. A boat's displacement is the dominant variable determining the size of wake, and hence the severity of erosive force along a river bank.

dissociation Many molecules in solution, and water itself, tend to break up (dissociate) into cations and anions. eg: $HCl \rightarrow H^+ + Cl^-$, or $H_2O \rightarrow H^+ + OH^-$.

dissolution The process of dissolving solids, such as the slow dissolution of rock minerals to produce ions in groundwater.

dissolved oxygen (DO) Natural surface waters are normally saturated with respect to oxygen. Polluted surface waters, hypolimnia and some groundwaters may be partially to totally depleted in oxygen. The saturated oxygen concentration, in $mg.l^{-1} = 14.16 - 0.3943T + 0.007714T^2 - 0.0000646T^3 - S(0.0841 - 0.00256T + 0.0000374T^2)$, where T is the temperature in °C, and S is the salinity in parts per thousand. Some quoted values are:

dissolved oxygen ($mg.l^{-1}$) in salt concentrations ($g.l^{-1}$)

T°C	0	7.5	15	22.5	30
0	14.6	13.8	13.0	12.1	11.3
10	11.3	10.7	10.1	9.6	9.0
20	9.2	8.7	8.3	7.9	7.4
30	7.6	7.2	6.9	6.6	6.3

Dissolved oxygen is best measured in-situ using a recently-calibrated specific ion electrode. Good maintenance of the electrode membrane is essential.

dissolved solids Ions and complexes in solution, as opposed to suspended solids. The size distinction between dissolved and suspended particles is arbitrarily set at 0.45 µm (an ultrafine filter pore size equivalent to about 1/200th the thickness of this paper). *See also* **total dissolved solids**.

distorted model Any model with unequal x - y - z scales.

distributary Divergent equivalent of a tributary: Any stream channel in a system of anastomosing or dividing channels, as in a braided stream, swamp or prograding delta.

distribution coefficient A partition ratio relating the adsorbed concentration of a groundwater pollutant to that of its dissolved concentration (assuming equilibrium); the slope of a linear adsorption isotherm.

divertible groundwater The long-term sustainable rate of groundwater extraction. *cf.* **safe yield**.

divining ... or water divining. *See* **dowsing**.

DNAPL Dense Non-Aqueous Phase Liquid = a 'sinker'. Any heavy low-solubility polluting liquid with a density > water, that sinks to the base of an aquifer, such as creosote, trichloroethylene, PCBs, halogenated alkanes, and a few unusual heavy oils. Short of digging out the aquifer, DNAPLs are notoriously difficult to get rid of, and often constitute a source of groundwater pollution for decades.

DOC Dissolved organic carbon. This is a generalized parameter to represent a huge range of natural and synthetic organics. Well over 2000 DOC compounds have been identified from drinking waters around the world, including carbohydrates, lipids, proteins, sugars, humic-, fulvic- and amino-acids, lignin and black carbon. Thousands more species have never been fully identified. The global flux of DOC in river waters is approaching a gigatonne per year. Some typical DOC values, in $mg.l^{-1}$, are:

Precipitation, and alpine rivers	< 1.5
Rivers in cold and temperate climates	3
Rivers in tropical rainforests	5 to 12
Leaf drip	5 to 10
Swamps, peat moors and wetlands	20 to 100
Oligotrophic lakes	1 to 3
Eutrophic lakes	2 to 5
Urban raw sewage	20 to 25
Random untreated wastewater	10 to 20
Treated sewage	4

Of themselves, dissolved organic carbons are not necessarily harmful, but such compounds can, and often do, form complexes with heavy metals, rendering them both more toxic, more soluble and more bio-available throughout the food chain.

doline = **vrtaca**. A depression, basin, 'O', 'D', 'T' or star shaped in plan, in which surface drainage recharges karstic aquifers. Dolines are often aligned along fractures, and are lined with clay, silt or other washed-in sediments which normally appear clearly on aerial photographs. They may vary from a metre to a km in diameter, and occur in densities of 0.5 to 250 km^{-2}. Dolines are amongst the most obvious and distinctive features of karst terrains.

dolomitisation The slow conversion of limestone to dolomite ($CaMg(CO_3)_2$) by the reflux of high Mg/Ca groundwaters. The process is complex, and is still not completely understood.

domepit A special case of a vadose shaft, found in karst limestones, and formed by the slow steady dissolution of groundwater dribbling down the shaft surfaces, thus creating solution flutes and grooves.

domestic water use The mean per capita urban water use, including garden irrigation and municipal water use, but *excluding* industrial water use and farming irrigation. Reasonable domestic water use in the developed world lies between about 150 and 350 litres per capita per day (depending upon climate). Extravagant domestic water use in some wealthy regions may exceed 1000 litres per capita per day. In many arid and semi-arid environments, such high consumption is becoming increasingly unsustainable. Domestic water use is sometimes, and incorrectly, thought of as 'drinking water'. In fact, drinking water seldom amounts to even 10% of domestic water usage.

Donor National and International organizations who finance most of the world's development projects. Donors tend to be as closely attuned to short-term political pressures as to long-term requirements in the water sector. Some have argued that, by fostering a culture of financial dependence, and by ineffectively tackling corruption, donors have done more harm than good. This is difficult to defend, but there is no denying that the success rate of donor-funded projects is dismal. Donors once 'copped flack' for funding inappropriate dam projects, but are now much more sensitive to issues of relocation, water management and climate change.

double mass curve analysis A graphical means of testing for homogeneity of data, or of compensating for systematic errors. The method plots two cumulative time-dependent data sets against each other. Any change in homogeneity resuts in a change of slope. The method has particular application in evaluating raingauge networks, and in comparing runoff from catchments under changing land-use conditions.

double porosity The void characteristics of an aquifer material that exhibits both intergranular (primary) and fracture (secondary) porosity. Aquifer tests in double porosity aquifers are amongst the most difficult to analyse.

down-hole TV Once considered a gimmick, but now developed to the point where a down-hole TV colour monitor can view a bore wall either sideways or axially. It is a useful technique for assessing fractured aquifers, and for maintenance and rehabilitation of old boreholes. Good illumination and no turbulence are required for success.

dowsing = **water divining**, = **water witching**. A supposed means of detecting groundwater (or other features) by means of 'extra-sensory perception' through a forked hazel stick, bent welding rods, or other non-scientific aids. It is an occult activity, which has been accepted as part of the 'New Age' philosophy. Even after centuries of practice, and innumerable controlled experiments, no scientific basis has been found, and it seems highly doubtful whether the results are any improvement upon chance alone. Dowsing is certainly a poor substitute for scientific hydrology but, curiously, it is still widely practiced within our western science-based culture. You are right to be highly skeptical of most hearsay on the subject!

drainage density Total drainage channel length per unit area (i.e. in km^{-1}). A geomorphic parameter sometimes used in catchment descriptions and empirical equations.

drainage divide = **watershed** (British usage).

drainage factor This parameter, B, is inversely proportional to the drainage time, in units of metres. $B = \sqrt{\dfrac{K_v.D}{\alpha.S_y}}$, where D is the thickness of the aquifer, K_v is its vertical hydraulic conductivity, $\dfrac{1}{\alpha}$ is the Boulton delay index, and S_y is the specific yield. For instantaneous drainage, $B = \infty$.

draft tube The wide-bore bell-mouthed exit tube from a turbine, designed to give minimum velocity head.

dragon's teeth Also known as **dreadnoughts** or **water baffles**. Concrete blocks, usually deployed in an array within a dam apron or tail race area to dissipate the kinetic energy of fast-flowing water, and thereby reduce scouring, bank collapse, *etc.*

drawdown The reduction in piezometric head due to pumping or gravitational drainage, especially relating to reservoirs and groundwater. In a pumped borehole the total drawdown $s = AQ + BQ^2$ where the AQ term is the formation loss, which is caused by Darcian flow, and the BQ^2 term is the turbulent well loss, or **excess drawdown** *(q.v.)*. A large drawdown can be caused by low transmissivity or low well efficiency, or both.

drill collar A heavy thickened drill rod between the bit and the rest of a drill string. The collar lends weight and rigidity to the drill string, and thus helps to keep the borehole straight and vertical.

drill string The kelly, and down hole parts of a drilling rig, comprising the bit, collar, and rods.

drilling fluid Usually a slurry of bentonite ± various additives. Alternatives include foam, water or air. The purpose of the drilling fluid is to flush out rock chips during drilling, to cool the bit, and to stabilise the open rockface by hydrostatic pressure.

drip irrigation = **trickle irrigation**. Irrigation water applied directly to a crop's root zone by means of leaky flexible pipes. The advantage is to minimize the applied water; the disadvantage is a requirement for continued labour-intensive maintenance.

dripstone Any rock deposited by dripping water, including tubes, stalagmites, stalactites, helictites, *etc.*

drizzle Light rain, strictly defined as <0.5 mm diameter raindrops falling with an intensity of <1 mm.hr[1].

drosometer An instrument for estimating dew-drop precipitation. A truly satisfactory drosometer has yet to be invented. Most existing designs have the usual problem that the instrument itself influences the parameter under investigation.

drowned flow Said of a weir or flume where the depth of water has increased to 'drown out' the hydraulic jump. An example is where the water depth is too great to sustain supercritical flow, and the unique stage-discharge relationship is lost. Flow can still be determined under drowned conditions, but requires readings from two stage recorders, instead of one.

drought A period of less than normal precipitation. There can be no rigorous definition of a drought because it is largely a function of perspective, context and geography. Some authorities distinguish between permanent, seasonal, contingent and invisible droughts.

dry deposition Atmospheric deposition of aerosols and suspended particles, and wet-surface scavenging of atmospheric gases, as opposed to deposition of the same from raindrops, snow *etc.* (wet deposition). Dry deposition is difficult to measure directly, but is at least of comparable importance to wet deposition in areas of industrial air pollution.

dry ice Solid CO_2, sometimes used to explosively blow clay, mud *etc.* from a clogged or undeveloped well. Also used in cloud-seeding experiments.

dry farming Non-irrigated farming. Especially crop growth dependent upon ephemeral rain and / or antecedent soil moisture storage.

dryland salinity Non-irrigated land with features indicating salinization (as opposed to irrigation or coastal salinization).

dryness index $D.I. = \dfrac{R}{L.P}$ where R is the radiation balance, P is the energy involved in evaporating any precipitation, and L is the latent heat of vaporization. The dryness index is a measure of aridity on the basis of an environment's energy balance. A D.I. of 1 to 2 is characteristic of steppe/prairie; 2 < D.I.< 3 is semi-arid, whilst a D.I. of > 3 is arid.

dry valley A valley or fluvial headwater, on porous rock, that seldom, if ever, contains surface flow. There are at least a dozen hypotheses for dry valley formation, most of which appeal to palaeoclimatic or extreme events.

DSS A decision support system is an interactive computer system to assist in management decisions in complex systems. DSSs are commonly linked to a GIS, computer models and database, and may be the key component of a Management Information System. DSS applications in the water sector include water resources or flood forecasts in complex river basins, dynamic rule curve analysis for multi-purpose reservoirs, multi-criteria analysis ('MCA'), and estuarine barrier management.

Dublin Principles In 1992 the 'Dublin Statement' on water and sustainable development was a PC response to the growing scarcity, pollution, mismanagement and imminent water-related conflicts of national water-sectors over much of the developing world. The main principles were: (1) *Fresh water is a finite and vulnerable resource, essential to sustain life, development and the environment.* (2) *Water* *development and management should be based on a participatory approach, involving users, planners and policy-makers at all levels.* (3) *Women play a central part in the provision, management and safeguarding of water.* (4) *Water has an economic value in all its competing uses and should be recognized as an economic good.*

Although these were all linked to an 'action agenda' which was laudable in principle, with hindsight these principles are now seen as naïve and insufficient. Inadequate international investment, inappropriate IWRM, poor technical backup, lack of follow-up, low political priorities, community indifference, and corruption, have all massively undermined the effectiveness of attempts to implement the Dublin Principles globally.

dug well — A shallow wide-bore well which is labour-rather than capital- intensive to produce. Where a shallow water-table is in danger of saline upconing, dug wells are much less risky than drilled wells. Countless dug wells have been abandoned worldwide, especially in India, due to falling water tables. That is, due to mismanagement of the water resource. Consequently, the rich, who can afford deeper drilled wells, get richer at the expense of the poor, who can only afford dug wells.

Dupuit assumption — Two simplifying assumptions, used in aquifer dynamics: that
(1) radial flow towards a pumped well in an unconfined aquifer is essentially horizontal, so the equi-potential lines are vertical, and (2) that the Darcian velocity is proportional to *tan i*, where i is the hydraulic gradient. Significant errors in the estimation of hydraulic conductivity, k, may accrue if the Dupuit assumptions are wrong.

duration curve A graph of river stage or discharge (linear or log 'y' axis) *vs* percentage time that a given stage or runoff is equalled or exceeded.

duricrust A broad spectrum of cemented horizons, most characteristic of the humid and arid tropics, comprising calcrete, silcrete and ferrocrete. Their formation mechanisms involve vadose groundwater. The hard pans so produced have an important influence upon surface flow, infiltration and groundwater occurrence.

Durov diagram Several variant graphical representations of some 8 or 9 hydro-chemical variables. An extension of a Piper diagram.

duty of water . . . or 'water duty'. Simply the amount of irrigation water per unit area; obviously, a function of climate, crop type, soil, irrigation efficiency, drainage and antecedent conditions. The term has become overused and imprecise, sometimes meaning water volume per application, area irrigated per unit volume of water, or total crop water requirement throughout the growing season. With such confusion, it is best to define your own meaning before using the term.

dyke (1) An impervious water-retaining embankment (of a river, lake or polder).
(2) Magma, generally basaltic, which has been injected into a fracture plane in a rock, and subsequently frozen into a sheet-like structure. Dykes are usually vertical, or nearly so, and of low (or zero) permeability. Consequently, they may form sub-surface barriers to groundwater flow.

dynamic equilibrium The general connotation may be used in any changing system. Alternatively, it may *specifically* imply a long-term state of balance between aquifer recharge and discharge, or, in a coastal environment, a stable saline interface against fresh-water discharge to sea.

dynamic ice Pressure due to a moving ice floe, over and above the static pressure. Dynamic ice raises river water levels, and hence increases bank storage. Note, in the context of climate change, dynamic ice refers not to rivers but to entire ice shelves and ice sheets.

dynamic source area Rainfall and recession alters water tables and soil processes. In consequence, the drainage network advances and retreats, and the catchment area actively involved in overland flow and streamflow, varies. This complex and dynamically interactive scenario supersedes older and simpler catchment models.

dynamic viscosity $\eta = \dfrac{F/A}{\partial v/\partial y}$, That is, the shearing force per unit area divided by the velocity gradient = coefficient of viscosity. A measure of the 'stickiness of a fluid'. *See* **viscosity**.

dynamic water level The water level in a well or borehole which is artificially lowered by pumping. *cf.* **S.W.L.**

dynamic watershed A conceptual watershed model in which the area of a catchment contributing to storm runoff is variable.

dystrophic lake A lake with a high content of externally derived organic matter.

E

earth dam *See* **embankment dam**.

Earth tide The slightly non-elastic response to gravitational cyclicity exhibited by the Earth's crust. Earth tides lag the oceanic tides by about an hour, and have a typical amplitude of about 10 to 30 cm (depending upon the 'local' geological elasticity). Earth tidal fluctuations of the piezometric surface, in confined and semi-confined aquifers, may have an amplitude of up to about 20 mm and this requires correction in aquifer tests. The effects of Earth tides must also be corrected in precise gravimetric surveys (such as to estimate the depth of alluvium prior to drilling). Earth tides can also influence microseismicity and volcanism.

ebb-and-flow spring = Intermittent spring. A rare kind of spring resulting from periodic siphoning from a sub-karstic reservoir. Episodic pulses of high discharge are superimposed upon a small, or even zero, baseflow. Fewer than 30 such springs have been described in the scientific literature.

Typical ebb and flow hydrograph

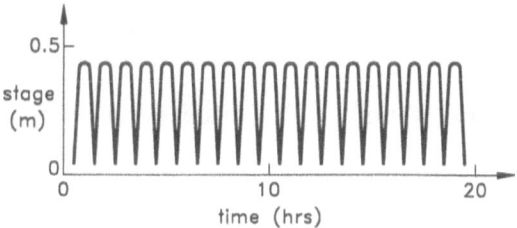

ebulition Boiling, bubbling, vigorous outgassing or effervescence.

eccentric well A well offset from its centre of influence, or from the centre of an elliptical or irregular area of influence. This occurs in an aquifer of variable hydraulic conductivity, or of markedly wedge-shaped section.

E. coli See **Escherichia coli**.

edaphic Of, or relating to, the soil. Edaphology: the study of soils in relation to water, the atmosphere, and plant growth.

eddy	Where a flow regime has an irregular boundary, (either solid or a chaotic contact between fluids of different velocity), then the quasi-circular swirls than develop are called eddies, or eddy currents. Eddies occur at all scales, from microscopic to major features of planetary ocean and atmospheric currents.

eddy diffusion	This is usually the dominant diffusive and dispersive process in flowing groundwater. Eddy diffusion, as distinct from molecular diffusion (a chemical process), is caused by inhomogeneity of flow through tortuous flow lines of varying pore diameters. In fine detail, several discrete processes operate, but all the various eddy-diffusive and molecular-diffusive processes can be lumped together and described by an overall diffusion coefficient.

eddy loss(es)	A 'bucket term' for various head losses in a pipe or channel system, for instance at valves, elbows, changes in diameter / width, in the lee of obstacles, scour pools *etc.*

edge contraction	The tendency of a nappe to pull away from the channel's edge as it accelerates in free fall.

edge exposure	Increased transpiration stress at the edge of a farm, woodland, orchard or other cropped area, resulting from both the increased exposed leaf area, and from greater advective drying.

EDTA	Ethylene-diamine-tetra-acetic acid. A selective complexing agent, once widely used to determine Ca^{2+} and $Mg^{2+}_{(aq)}$. The method involves removal of interfering ions, colourimetric titration with EDTA at pH > 12 to give Ca^{2+}, with subsequent titration at lower pH to give Mg^{2+} This is now largely superseded by AAS and ICPAES.

Dictionary of Hydrology and Water Resources

eductor pipe A pipe containing flowing water and entrained air as, for example, in the 'air-lift' development of a well.

eductor pump A simple but crude and inefficient form of pump in which water, accelerated through a constriction (venturi flow), lowers the pressure at an inlet point, thereby creating suction.

effective diameter = **effective size**.

effective fetch The mean fetch throughout the direction of prevailing wind ± 45°.

effective flood control storage (1) *of a reach of a river* ... The difference between baseflow volume and floodpeak (overbank) volume.
(2) *of a reservoir* ... The difference between flood peak volume and dead storage volume.

effective head In hydro-electric power, that proportion of head potential which is actually converted to electrical energy, *not* including friction and kinetic losses.

effective porosity ϕ_c = the volume of groundwater that can drain gravitationally ÷ the volume of rock containing the groundwater. This is strongly influenced by the particle size. Some comparisons between total and effective porosity (after Chapellier, 1992) are:

rock type	total porosity %	effective porosity %
gravel	45	40
sand	40	30
silt	36	3
clay	47	0
chalk	30	2 to 5
fissured crystalline	0.1 to 2	0.1 to 2

effective pressure   The intergranular pressure of an aquifer (as opposed to the hydrostatic pressure).

effective rainfall   (1) *Hydrology:* that fraction of rainfall that contributes to surface runoff (quickflow).
(2) *Irrigation:* that proportion of rainfall that is held by the soil, and is available for plant growth.
(3) In basin modelling, the 'effective rainfall' is a weighted mean of current and antecedent rainfall which best correlates with the runoff, usually on a daily basis.

effective size   Whatever *you define it* to mean! There are at least six mutually inconsistent definitions in common usage; generally of the form *"That particle size of a soil or sediment sample at which x weight % is finer"*. See *also* **equivalent size**.

effective storage   = **effective capacity**. The usable storage of a reservoir, between spillway and lowest intake. The effective storage decreases slightly with time, due to sedimentation. *cf.* **peak storage**.

effective velocity   An estimate of the true mean velocity of groundwater, given by the Darcian velocity ÷ porosity.

efflorescence   The process in which many hydrous salts, and especially many sulphates, undergo subaerial dehydration (partial or total), thereby yielding a bleached-looking and often powdery crust. The opposite of deliquescence.

effluent   (1) Suspended and dissolved waste products (such as from urban runoff, sewage and industrial wastes) carried by, and hence polluting the water.
(2) An effluent condition is one in which there is a nett loss of stream water to groundwater by

Dictionary of Hydrology and Water Resources

seepage through the stream bed. An effluent stream = a gaining stream.

EIS A legally required document to elucidate the environmental impact of a project. Environmental impact statements are frequently prepared by interested parties. Hence they may vary from impartial scientific studies to cynical exercises in political deceit.

electrical conductivity = **specific electrical conductance**, specific conductance, or just 'conductance'. EC used to be reported in micromhos (μmhos) but is now redesignated as microsiemens (μS). Strictly, one should quote EC as μS.cm^{-1} at 25°C, but most authors take the 'cm^{-1} at 25°C' as read. In some respects EC is a more significant measure of salinity than **TDS** *(q.v.)*. It is also much quicker and easier to measure. *See also* **TDS meter**. The electrical conductance *vs.* salinity relationship is linear up to roughly 5000 μS. That is up to slightly brackish conditions.

electrical resistance model Electrical analogues of groundwater flow, from the pre-digital computer era. Wonderfully impressive to look at, and great fun to use, but nowhere near as adaptable as numerical models. Every science museum should have one!

electric log = **E log** ≈ **resistivity log**. A combination of electrical resistance log and self potential log. Perhaps the most useful and commonly used of all the down-hole wire-line probes for investigating the groundwater potential of recently drilled boreholes.

electrolysis . . . *of water* . . . Passage of an electrical current through water, sufficient to obtain gaseous hydrogen at the anode, and oxygen at the cathode.

electro-osmosis The effect of increased permeability in a clay or comparable substance, caused by applying an electrical potential. Typical values range from 2-12 x $10^{-9} m^2.V^{-1}.s^{-1}$.

elevation of prime E_p. The maximum elevation at which snowmelt can occur under given seasonal conditions. E_p is a function of: snow water content, seasonality (such as number of days since mid-winter), and the decayed annual temperature (such as the degree days at 1900 metres since January 1st (northern hemisphere) calculated with a decay constant of 0.96). E_p is an important parameter in some snowmelt models.

eluviation The removal of *soil* particles in suspension.

elutriation The removal of any particulate matter from clear water by decanting or filtering.

embankment dam Alias **earth dam**. A range of dam designs based upon the central theme of an elongated mound, in which the hydrostatic pressure of the reservoir combines with gravity to maintain a stable configuration. The two basic variants are earth-fill and rock-fill. It is common to have an impermeable core comprising clay, asphalt, concrete or synthetic impervious membrane. A drainage system is used to drain seepage at the toe. An embankment dam should never overspill unless it incorporates an adequately designed concrete overspill section. For large embankment dams it is essential to pay particular attention to the foundation geology, including seismicity. Seepage under the dam is a major cause of dam collapse. There is no statistical information on the number of embankment dams in the world.

embedded water Water used to grow food and make things. Essentially the same as 'virtual water', but embedded water is used in the context of the real water used in the production of individual items, whereas virtual water is used in the context of water exportation, generally international, and commonly from water poor to water-rich countries. Examples of e.w. are: One pint of beer embeds about 75 litres of water. One cup of coffee embeds 140 litres of water. One T-shirt embeds 2000 litres of water. Such estimates incur huge margins of error and a host of implicit assumptions, so should not be taken too literally. There are several philosophical problems with the concept of embedded water, such as the assumption that all sources of water are of equal financial or environmental value, or are equally fungible.

emulsion A suspension of one liquid within another as a homogeneous distribution of micro-droplets, such as milk.

end contraction If a sharp-crested weir does not span the full channel width, then at the weir, a component of flow exists away from the stream banks (perpendicular to the main channel flow), which causes the nappe to contract. This effect requires an empirical correction in discharge formulae.

endorheic Adjective describing a 'blind drainage' system; an internal drainage basin such as the Caspian and Aral Seas, Lake Eyre, Great Salt Lake *etc.*

endokarst Underground karstic features such as caves and underground lakes, dripstone, dissolution pipes, crossovers, siphons, *etc.*

end point A stage of a reaction indicated electrochemically or colorimetrically, which marks the end of a titration; the point of inflexion on a titration curve. For example, titrating a carbonate-bicarbonate sample with acid will give a curve with two inflexion points. The curve should be plotted by readings from a specific ion electrode (pH in the carbonate-bicarbonate example), but the end points are often approximated by a colour, such as the methyl orange end point of an alkalinity titration. *See also* **P and M end points**, and **indicators**.

energy coefficient = **Energy correction factor 'α'**. *See* **kinetic energy correction factor**.

energy grade line The notional profile of the sum of the potential, pressure and kinetic energy potentials along a flowpath. It is almost the total energy line, but does not include frictional head losses.

energy gradient The local gradient of the energy grade line. In a slow moving river, the mid-channel water surface is a close approximation to the energy gradient.

englacial stream A stream that flows *within* glacial ice. Flow is more commonly subglacial.

English rule A rule that any landowner has an absolute right of ownership of any groundwater beneath his or her land. In these more enlightened times, the English rule is only tenable if a *responsible* landowner owns the *entire* catchment. The English rule is essentially incompatible with modern philosophies of total catchment management.

Entamoeba *Entamoeba hystolica* is an aquatic parasitic protozoan which causes abdominal discomfort to severe dysentery, and which is occasionally fatal.

enteric micro-organisms The *Enterobacteriaceae* is a large group of Gram-negative, rod-shaped anaerobic bacteria, mainly found in the gut of animals, and including

Escherichia coli and some pathogens. Enteric infection can occur through drinking water. *See* **protozoans**.

environmental flow The minimum discharge of a river required to maintain its fresh-water and estuarine ecosystems and to conserve human livelihoods dependent upon those ecosystems. In principle the water resource availability of a river is the difference between 'natural flow' and the 'environmental flow'. In practice it proves impossible to quantify this flow to the satisfaction of all stakeholders. Some environmentalists argue that *any* water extraction from a river is to the detriment of its ecosystem. How does one determine the environmental flow? 20% of the total? 10%? 5%? The crunch comes during times of drought. When the riverine ecosystem is most stressed is also the time when the demand for human water supply is greatest. In most countries the political imperative to maintain water supplies generally trumps environmental considerations.

environmental isotopes Both natural and artificial isotopes (stable or radioactive) that are widely dispersed throughout the environment. In hydrology the environmental isotopes of most interest are those of hydrogen, oxygen and carbon, but a wide range of synthetic radioactive tracers can also be used in circumstances where they rapidly decay to extinction before human contact.

environmental pathway The route by which a radionuclide or toxic pollutant can actually or potentially reach humans, such as by biological concentration in edible aquatic organisms.

ephemeral Short-lived, such as an ephemeral spring or flash flood. An ephemeral stream is also known as an intermittent stream.

epikarst The surface and near-surface (subcutaneous) zone of a karstified formation, where dissolution has been intense. Epikarst is not quite synonymous with exokarst, the latter referring to surface features only.

epilimnion A low density, and hence uppermost, layer of a stratified water body. Generally the most oxygen-rich layer, with the best quality water and most diverse aquatic ecosystem. Most sunlight is available in the epilimnion, and hence this layer contains the richest phytoplankton.

epitaxial growth A form of cementation in which the cementing mineral is precipitated in crystallographic continuity with crystals in the original rock grain.

epitaxial occlusion A form of crystal growth to the extent that all pores are filled with crystalline precipitates, and the rock becomes impervious.

EPM Equivalent Porous Medium. An anisotropic aquifer, possibly of double or triple porosity, which may, for some purposes and on a sufficiently large scale, be conceptualized as an equivalent homogeneous matrix of transmissivity $\overline{T} = \sqrt{T_{max}.T_{min}}$

Epper effect The problem that a stream gauge, when immersed in the stream to measure point velocity, itself affects the velocity being measured. The Epper effect is particularly important near the lower threshold of sensitivity.

equal area slope For a given catchment, the thalweg is drawn, and a constant slope superimposed such that area A = area

B. This representative gradient is used in various empirical formulae, such as the derivation of a synthetic unit hydrograph.

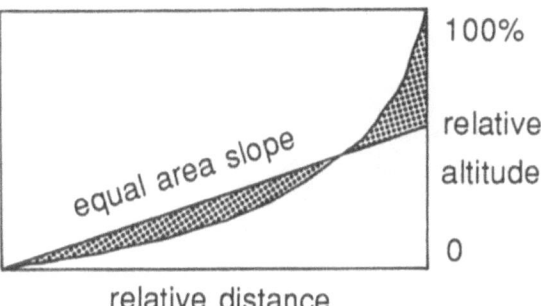

equilibrium line .... *of a glacier* ... The local elevation at which annual ablation just balances the annual snow accretion.

equilibrium price The commodity price, P_e, at which supply just equals demand.

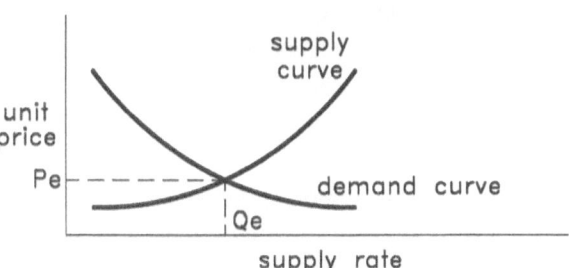

equinoctial rains The annual oscillation of the ITCZ, approximately between the tropics, tends to bring biannual rains to the equator at about the spring and autumn equinoxes.

equivalence Ambiguity in terms of a trade-off between layer thickness and apparent resistivity in a vertical electrical sounding.

equivalent salinity The sodium chloride concentration that would produce the same E.C. as the mixed solute water under consideration.

equivalent size (diameter) That uniform spherical grain size (or diameter) that would transmit water at the same rate as a natural soil / sediment / rock sample of heterogeneous grain size and shape; the synthetic grain size at which $K_{synth.} = K_{sample}$. Sometimes incorrectly termed 'effective size'.

equivalent weight = **Combining weight**, = the weight of a substance (in grams) that will react with 1 gram of hydrogen or 8 grams of oxygen. It is determined as: *the atomic or molecular weight ÷ magnitude of its ionic charge*. Strictly, it could also be defined as: *the molecular weight ÷ the oxidation number change in reaction*, in which case it depends upon the reaction being considered. However, in most cases the equivalent weight is unambiguous.

equipotential line A line of equal pressure/head/potential, somewhat analogous to the 'strike' in solid geology. *cf.* **flow line**.

equivalent The number of equivalents in a solution = the concentration in $g.l^{-1}$ x nett positive valency ÷ molecular weight. In most water analyses it is convenient to use milliequivalents ('$meq.l^{-1}$').

error assessment In 40+ years as a water resources consultant, in 23 countries, this author has yet to see a water resources report that was published with a satisfactory error assessment. That is, (almost?) the worldwide water industry is guilty of unscientific and unprofessional publication of data, without adequate error bounds and without realistic implied precision. For example, quoting a basin water resource as '8,745,760 m^3/year' instead of '8.8 ± 2.3' $Mm^3.yr^{-1}$, is bordering

Dictionary of Hydrology and Water Resources

upon scientific fraud! In particular, the final derived value cannot be more precise that the original measured values. Rainfall and river flow is commonly measured to a precision of ± 5 to 10%. Evaporative and transpiration loss, infiltration and aquifer dimensions are generally even less precise. Reports which imply required hydrologic precisions of anything from 4 to 8 significant figures are absurd and misleading. Moreover, it is unfair to present this to a non-technical client who may well believe data to such high precisions.

error ass't cont Throughout most of the world it proves inordinately difficult to persuade 'professionals' to take error assessment seriously. If a water manager can't undertake realistic error assessment, then he or she is incapable of doing the job.

error (rainfall) Quite apart from point measurement errors there is a spatial error, which is inversely proportional to the rainfall amount and network density, but proportional to the sampling time-interval, Δt. An empirical 'rule of thumb' which seems to work quite well for non-mountainous regions of Australia (and elsewhere?) is:

$$E = 8500 \frac{\Delta t}{R^{0.6} . A^{0.5}}$$

Where E is the RMS error, as a % of the monthly rainfall, R is the mean areal rainfall in mm, A is the mean area represented by each gauge in the network, and Δt is the sampling time interval in hours.

error theory There are two types of error, random and systematic. *Random* error is generally caused by human mistake, and is inherently unquantifiable. *Systematic* error occurs with *all* measurements. Suppose U is a function of independently measured / estimated variables; that is, $U = f(A, B, C, ...)$ where the respective measurement errors are $\Delta A, \Delta B, \Delta C, ...$

etc., then according to error theory, the uncertainty, can be expressed in such rules as:

If $U = \pm A \pm B \pm C ...$,

then $\Delta U = \sqrt{(\Delta A)^2 + (\Delta B)^2 + (\Delta C)^2 + ...}$,

If $U = \dfrac{A.B...}{X.Y...}$,

then $\Delta U = \sqrt{\left(\dfrac{\Delta A}{A}\right)^2 + \left(\dfrac{\Delta B}{B}\right)^2 + ... + \left(\dfrac{\Delta X}{X}\right)^2 + \left(\dfrac{\Delta Y}{Y}\right)^2 + ...}$,

and if $U = kA^\alpha$, then $\Delta U = k \propto A^{\alpha-1} \Delta A$

'ΔU', is the uncertainty, or error, and may be presented in either absolute or percentage terms. The above, and related statistical rules, are foundational to the professional analysis of water resources. Caution: pay attention to consistency of units.

Escherichia coli   An index bacterium whose presence in water is indicative of faecal pollution. Hence *E.coli*, in general, is universally used as the primary test for biological purity of water supplies, whence its presence indicates a risk of less obvious but possibly pathogenic bacteria. The only problem with *E.coli* tests is their slowness. With standard equipment a presumptive test takes 18 hours, and confirmation takes 72 hours. Most *E.coli* infection, both pathogenic and otherwise, occurs not through water, but from meat prepared under poor hygienic conditions. *E.coli* is normally harmless, and is benignly present in the human bowels, but five of the 190 known serotypes can cause urinary-tract infections, gastroenteritis, anaemia and haemolitic-uremic syndrome. Serotype *E.coli* O157:H7 may even be fatal

esker   Sub-glacial fluvial sediments. Eskers are characteristically sinuous linear deposits.

estavelle A reversible spring found in some karst areas. It is a spring in the wet season, but a swallow hole in the dry season.

Euler's equation This is a widely used force balance equation for variations of pressure, density and elevation for fluid flow along a streamline. It is based upon Newton's second law of motion. The simpler version applies only to *steady* flow. There are many versions of this equation, such as:

$$\rho\left(\frac{\partial v}{\partial t}+v\frac{\partial v}{\partial s}\right)+\frac{\partial P}{\partial s}+\rho g\frac{dg}{ds}=0.$$

euphotic zone (Grk: eu= well or richly, photic, of light) The water layer which extends from the surface to a depth at which 99% of the light is absorbed. This varies from a few mm to about 40 metres.

eutrophic(ation) (Greek: trophos = nutrition) Waters rich in nitrates and phosphates which develop excessive growth of planktonic blue-green algae. This in turn depletes the hypolimnetic oxygen, resulting in fish deaths and reduced stream-bed conditions. The algal blooms are often smelly or even poisonous. Eutrophic waters are usually caused by pollution, but some slow moving lakes, creeks or ox-bows eutrophy by natural means. *cf.* **oligotrophic** and **mesotrophic**.

evaporation line Evaporation of near-surface groundwaters, or surface waters, causes a progressive enrichment of heavy isotopes in the residual water. When plotted on a $\delta^{18}O/\delta D$ diagram this isotopic evolution follows an evaporation line of the form: $\delta D = a.\delta^{18}O + b$, where a normally lies between 4 and 8, and b <10.

evaporation pan A shallow tank of water, with an accurate means of measuring the water level, used to estimate actual daily evaporation from an open water body (≈potential evaporation from a saturated soil). Dimensions of the 'American Class A Pan' are 122 cm diameter, 25 cm deep, with a rim 10 cm above ground level, and a water level not less than 10 cm below the rim. The class A pan is now almost universally standard, although both larger and smaller pans are occasionally used. The larger the pan, the more accurate the reading. Errors arise from heat conduction through the base, atypical or extreme pan exposure, and birds drinking or bathing in the water. *See also* **pan coefficient**.

evaporativity   The potential rate of evaporation under stated conditions.

evaporimeter   An instrument for measuring the evaporation of water, or a saturated surface, into the atmosphere. The most common evaporimeter is a Class A evaporation pan with micrometer hook gauge and bird screen. Others include the now obsolete atmometer, and Piché gauge.

evaporite   A class of sedimentary rocks formed either by evaporative concentration of brines or by brine interaction with pre-existing evaporite minerals. The vast majority of evaporites consist of halite, gypsum and/or anhydrite. Of all rock types, evaporites have the lowest density, highest solubility, and are most susceptible to plastic deformation. Some evaporite-carbonate rocks are high yielding aquifers, but give poor quality waters. There are scores of evaporite minerals, such as:

The more common evaporitic minerals

aragonite	$CaCO_3$	calcite	$CaCO_3$
dolomite	$CaMg(CO_3)_2$	magnesite	$MgCO_3$
gypsum	$CaSO_4.2H_2O$	halite	$NaCl$
mirabilite	$Na_2SO_4.10H_2O$	natron	$Na_2CO_3.10H_2O$
sylvite	KCl	rona	$Na_3CO_3.HCO_3.2H_2O$

some less common evaporite minerals (☛ typical salt-lake zeolites)

thenardite	Na_2SO_4 $Na_2CO_3 \cdot CaCO_3 \cdot 5H_2O$	gaylussite	
epsomite	$MgSO_4 \cdot 7H_2O$	hexahydrite	$MgSO_4 \cdot 6H_2O$
celestite	$SrSO_4$	huntite	$Mg_3Ca(CO_3)_4$
kieserite	$MgSO_4 \cdot H_2O$	tachyhydrite	$CaCl_2 \cdot 2MgCl_2 \cdot 12H_2O$
burkeite	$Na_6(CO_3)(SO_4)_2$	bloedite	$Na_2SO_4 \cdot MgSO_4 \cdot 4H_2O$
antarcticite	$CaCl_2 \cdot 6H_2O$	glauberite	$Na_2SO_4 \cdot CaSO_4$
thermonatrite	$NaCO_3 \cdot H_2O$ $Na_2CO_3 \cdot CaCO_3 \cdot 2H_2O$	pirssonite	
nahcolite	$NaHCO_3$	erionite	$NaKAl_2Si_7O_{18} \cdot 6H_2O$
carnalite	$MgCl_2 \cdot KCl \cdot 6H_2O$	clinoptilolite	☛$NaAlSi_5O_{12} \cdot 4H_2O$
bischofite	$MgCl_2 \cdot 6H_2O$	magadiite	☛$NaSi_7O_{13}(OH)_3 \cdot 3H_2O$
nesquehonite	$MgCO_3 \cdot 3H_2O$	artinite	$MgCO_3 \cdot Mg(OH)_2 \cdot 3H_2O$
sepiolite	$Mg_2Si_3O_6(OH)_4$		
hydromagnesite	$4MgCO_3 \cdot Mg(OH)_2 \cdot 4H_2O$		

evapotranspiration 'ET' is the combined water loss by evaporation from the ground or flooded paddy, and transpiration by vegetation. The actual evapotranspiration from a cropped or naturally vegetated area is generally somewhat less than 'reference evapotranspiration', ET_0, as either measured or calculated (for example, by the Penman-Monteith equation). This discrepancy arises from soil-moisture constraints, crop physiology, and air-frictional loss of advection. Hence the actual crop water use is estimated as the product of ET_0 and *evapotransp'n cont.* the 'crop coefficient', K_c. The latter may be obtained from empirical tables published by the FAO, *et al.* In the temperate regions of Europe and North America about 60% of the precipitation is directly lost to evapotranspiration—much more in the tropics and much less in polar regions.

evorsion Erosion in a river bed caused by turbulent flow.

exceedance *See* **AEP**.

excess drawdown The drawdown in a pumped well or borehole has two components: the normal aquifer response (hopefully > 90% of the total drawdown), and excess drawdown. These two components may be of comparable magnitude in a poorly constructed well *(see* **well efficiency**). Factors influencing the excess include: turbulent inflow within or close to the well (the main cause), poor well development, insufficient open area of the screen, insufficient length of screen, wellscreen corrosion / encrustation, an unsuitable filter pack, and / or significant vertical flow in the formation ($K_v < K_h$).

excess rainfall Or excess precipitation. The nett precipitation after interception and evaporation. The fraction of precipitation contributing to quickflow or groundwater recharge.

exchange capacity Normally a contraction of **Cation exchange capacity** *(q.v.)*, although anionic exchange can occur under acidic conditions and on some mineral surfaces.

exhaustion The sediment load of a stream or river is controlled by fluvial dynamics, and by the availability of sediment within the fluvial system. Usually, the former is the limiting factor, but sometimes the sediment load

Dictionary of Hydrology and Water Resources

decreases during a storm because of exhaustion—a lack of sediment replenishment in the source area.

exit point   The transition point on any hillslope, between infiltration and 'exfiltration'.

exokarst   Surface characteristics of a karstic terrain, such as dry valleys, dolines, swallow holes, rillenkarren and stream resurgences.

exorheic   Converse of endorheic; any basin with drainage to some external point (usually the sea).

expert system   Computer software capable of assessing complex multi-parameter systems, and capable of 'learning and adapting' from case histories.

exploitable yield   The theoretically maximum achievable water yield, usually of a catchment. In most catchments, to fully achieve the exploitable yield would be ecologically and aesthetically disastrous: it should be regarded as a theoretical value, not an objective!

explosive development   A last-ditch, and fairly desperate attempt to improve the yield of a well by explosive fracturing of the aquifer. The process is spectacular, but the chances of improving the well are about equal to the chances of destroying it!

exsurgence   A karstic spring apparently unrelated to an upstream swallet, or any spring outflow of autochthonous groundwater. Arguably a pedantic distinction from a resurgence.

extinction coefficient   For a given wavelength, vertical light penetration into a water body decreases exponentially with depth according to the expression: $I_z = I_s . e^{\varepsilon \lambda . z}$, where I_z is the intensity at depth z, I_s is the surface light intensity, and ε_λ is the extinction coefficient.

extinction depth   The depth of a water table below the surface at which evapotranspiration does not occur. This is effectively about 2 metres for evaporative soil moisture loss by capillary rise, but may be tens of metres for transpiration by deep-rooted phreatophytes. Strictly, the moisture loss decreases exponentially, and hence the extinction depth cannot be accurately defined.

eyot   A small island in a lake or river.

facies   Most commonly, a suite of rock types derived from a particular environment or set of formation conditions. Sometimes also used in relation to water chemistry, such as 'a Ca-HCO_3-SO_4 facies brine'. In general, a grouping with connotations of similar origin.

factor analysis   A technique of statistical analysis used to determine the relative influence of multiple variables.

facultative   Descriptive of an organism that can grow under various conditions, as opposed to being obligate, such as a facultative aerobe.

faecal pollution   = **fecal pollution**. Water which has at some point come into contact with sewage, either from human or animal waste, and in consequence of which, may contain enteric pathogens.

falaj   = **falag**. *See* **qanat**.

fast neutrons — Neutrons at energies of 4 to 6 MeV, such as emitted by ^{226}Ra-Be, ^{239}Pu-Be, or ^{241}Am-Be, which are used in borehole logging or soil moisture profiling.

fault spring — A structurally controlled spring, or spring line, in which an aquifer discharges against a fault-controlled impervious threshold.

Fen — See **marsh**, *swamp*, *cf.* **bog**.

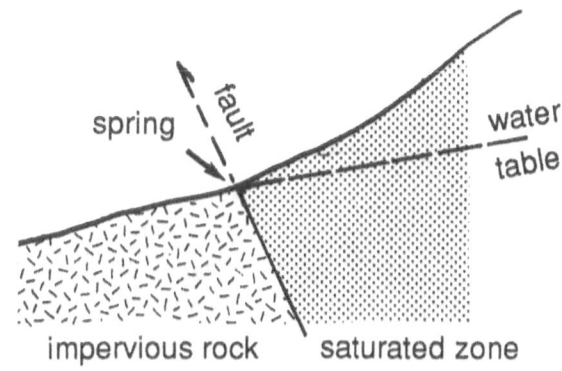

fenglin — (Chinese) 'Stone forest'. A tower karst landform consisting of isolated pinnacles. An advanced and somewhat spectacular product of karstic dissolution.

fengcong — (Chinese) A tower karst landform in which multiple pinnacles rise from an exposed rock base.

fermentation — The bacterially catalyzed degradation of complex organic molecules to produce energy and simpler molecules, resulting in a more soluble organic load. Reactions take the form: $A \rightarrow A' + CO_2$. eg: $2C_{organic} + H_2O \rightarrow CO_2 + CH_4$. There are numerous such reactions involved in biological decay, such as rotting leaf litter in soils, or reducing organic sediments in lakes.

ferralitisation — Intense leaching under tropical weathering conditions to leave a kaolin residual soil.

fertigation	The delivery of soluble fertilizer through irrigation water. With careful calibration of crop nutrient requirements, combined with carefully controlled delivery, such as drip irrigation, fertigation can be the most efficient means of crop optimization.
fetch	The maximum straight line distance across a water body in the direction of the prevailing wind.
Fick's laws	These two laws sum up everything that is important about diffusion. i) $J = -D\dfrac{\partial c}{\partial x}$ where J is a solute flux, C is the solute concentration, and D is the diffusion coefficient (in $m^2.s^{-1}$). *Note:* the solute flux is proportional to the concentration gradient. The negative sign shows that the flux occurs down-gradient.
Fick's Laws cont.	ii) $\dfrac{\partial C}{\partial t} = D\left(\dfrac{\partial^2 C}{\partial x^2}\right)$, that is, the concentration gradient is a function of the rate of change of concentration with respect to time. Thus diffusion alone is never able to achieve perfect homogenization of a solute. (Similar equations can be derived for heat flow).
field blank	Pure water which is processed through the normal sampling and analytical procedures to determine whether contamination or systematic errors are occurring. This is an important test when dealing with very dilute waters such as rain water.
field capacity	= **field moisture capacity**. The residual water held in a freely draining soil after some given period (usually a day or two) of drainage from initial saturated conditions. Hence, the volumetric water

Dictionary of Hydrology and Water Resources

content in the vadose zone after the gravitational water has drained to the water table.

field coefficient ... *of permeability* ... The coefficient of permeability at ambient temperature.

field deficiency = **field moisture deficiency**. The water required to restore a soil to field capacity.

field duty *of water* ≈ Farm duty of water. The irrigation water delivered per unit area of agricultural land per year, or per growing season. Units of $m^3.km^{-1}(yr^{-1})$ or $kL.ha^{-1}$.

field waste Surface runoff from an irrigated area. If the crop is irrigated efficiently, there shouldn't be any field waste. Laser levelling of the cropped area, prior to planting, is a good way to reduce field waste.

film flow = **Capillary flow**, = **pellicular flow**. Flow (resulting from gravity and osmosis) within the thin film of water adhering to particles in the unsaturated zone.

filter cake = **mud cake**. When drilling a borehole, the hydrostatic pressure in the bore forces some of the drilling fluid into the adjacent pores of the aquifer. Thus clay particles are filtered out against the side of the bore, thereby forming a tubular-shaped filter cake. This inhibits groundwater flow, and has to be removed prior to production pumping. *cf.* **filtrate.**

filter crop Reeds, papyrus, water hyacinth, cress *etc*, grown in a channel to trap sediment particles; a kind of artificial wetland. Filter crops may also be grown in strips on a hillside to inhibit soil erosion.

filter pack The tubular space between the drilled wall of a borehole and the inserted casing / screen is commonly filled with a filter pack consisting of uniformly-sized gravel. This has the advantages of preventing formation collapse, increasing the efficiency of the well, and decreasing siltation within the well.

filtrate In general, any filtered fluid. In borehole drilling operations, the drilling mud is hydrostatically forced into pores of the surrounding rock. This filtering action causes the mud to partially separate into the filter cake (a layer of clay particles encrusted onto the bore wall), and filtrate (a watery fraction of the drilling mud which invades an annular space around the bore).

filtration There are numerous methods of filtering a water sample or supply, the purpose being to remove suspended solids, including the larger biota. For analytical purposes the filter's pore size is conventionally 0.45 mm. Normal filtration *alone* removes neither dissolved salts nor the smaller pathogenic organisms (such as some viruses).

fines The fraction of a soil or friable rock that has a grain size less than some arbitrary limit. Sediment that takes a long time to settle out of suspension.

finite difference model A numerical model designed to calculate the configuration of some complex variable, such as hydraulic potential, within described boundaries, according to defined initial and boundary conditions. The analysis is iterative and based upon the assumption that the hydraulic properties of an area may be approximated by point values distributed on a regular rectangular grid.

finite element model Similar to a finite difference model, except that (1) The grid consists of irregular triangles, and (2) the working variable is calculated as a surface rather than as a point value. Finite element theory is more complex than finite difference theory, but the method is more flexible in that the cells can be of varying size and shape to suit the complex geometry of real aquifers. Smaller triangles can also be defined for areas of particular interest or complexity. Another difference between finite difference and finite element models is that interpolation functions can be used within the grid triangles of finite elements to determine the head at any point, whereas with finite difference grids, the head is only defined at nodal points.

fingering When a high density liquid overlies a less dense liquid, or when a fluid spills onto an unsaturated granular medium, the resultant vertical flow is concentrated into discrete flow-paths (fingers) of very complex geometry. This fingering is, to some extent chaotic, and cannot be exactly reproduced by either digital or analogue simulations.

firn = **névé**. A transitional texture between snow and ice, characterized by snow compaction and freeze-thaw or pressure metamorphism to a granular, and ultimately, to a massive texture.

firn line The boundary between fresh snow and firn in the upper section of a glacier.

first flush The first pulse of runoff from snowmelt after the winter freeze. The first flush carries most of the contaminants from the snowpack, and some of the soil leachate from the winter soil-moisture residence. Consequently it is chemically distinct, and sometimes remarkably acidic.

fish(ing) The frustrating process of trying to recover a broken bit, logging tool or jammed pump from down a borehole. Despite great ingenuity in the design of fishing tools, some lost equipment is just not recoverable. Well-designed logging probes should break at a pre-determined 'weakest point', which can be easily 'grabbed' by a fishing tool. *See also* **zero defect**.

fish ladder = **fish bypass**. A stepped series of pools and waterfalls to enable migrating fish to 'swim around' obstacles like dams, in order to spawn upstream.

flash flood A high intensity, short duration flood, typical of small and/or steep catchments and of arid, urban and deforested environments.

flaw A short sharp gust of wind.

floating pan An anchored floating evaporation pan used on lakes. It has a surrounding wave deflector or dissipation ring, and provides some of the best field estimates of open water evaporation.

float well Any stilling well in which the static water level is measured by a float gauge.

floc An aggregate of clumped particles, commonly clays or organics, which is held together by Van der Waals forces. This is typical of estuaries where dilute water meets sea water: the increased ionic strength collapses the electrical double layer which would otherwise repel particles electrostatically. Flocculation agents may also be added to water during various purification processes.

flood coefficient = **the runoff coefficient**.

flood envelope In the absence of any gauged data, regional flood envelopes may be a useful first approximation of expected flood maxima. The highest recorded specific discharge ($m^3.s^{-1}km^2$), or peak discharge ($m^3.s^{-1}$), is plotted against drainage area, in km^2. Commonly both axes are logarithmic which may yield the illusion of a linear relationship, but there is no theoretical reason why log Q vs log A should be linear. Indeed, some curvature is to be expected. It is common practice to present a family of curves, which can give estimates for different regions for the same ARI, or for a single region with different ARIs. Or the envelope can be quoted in the form $Q_{ARI} = aA^b$ where a is a regional statistical coefficient based upon climate and geomorphology, A is the catchment area, and b is an exponent of <1. If the ARI is not given, then Q is implied to mean the probable maximum flood. Some random quoted examples are:

$$\text{Global Maximum } Q_{max} = 154A^{0.738}$$
$$\text{Typhoon south and southeast Asia } Q_{max} = 850A^{0.357}$$
$$\text{Flinders Ranges, South Australia } Q_{max} = 10.9A^{0.71}$$
$$\text{New Zealand } Q_{mean} = 110A^{0.82}$$
$$\text{Average of 9 Japanese rivers } Q_{mean} = 50A^{0.5}$$
$$\text{Variants include Creager's formula: } Q_{max} = CA^{bA-n}$$

flood irrigation   A method of irrigation in which an entire field is flooded. Even with laser levelling, this is an inefficient method of irrigation.

flood peak   Usually, the maximum discharge rate of a flood, in cumecs, but in the context of flood *volumes*, the flood peak is that part of the hydrograph between rising and falling points of inflexion.

flood plain   A flat riverside area immersed during a major flood event. Flood plains are usually highly fertile due to a shallow water table and the deposition of nutrient-rich sediments.

flood risk This is becoming increasingly difficult to quantify. The risk is greatest in coastal flood-plain cities susceptible to the 'quadruple-whammy' of sea level rise, tidal surge, climate-change-linked increases in runoff intensity, and hydro-compaction of underlying aquicludes. As a percentage of GDP, the five most vulnerable cities, as of 2014, are: Guangzhou, New Orleans, Guayaquil, Ho Chi Minh City (=Saigon), and Abidjan. But thousands of other cities and towns are at serious risk, including many that, in the past, have not been regarded as being 'high risks'.

flood routing *See* **routing**.

flood storage Reservoir capacity designed for attenuating floods. The associated stage may be above the spillway level.

floodway = **by-pass channel**, = flood relief channel. A conduit designed to carry a flood peak out of harm's way.

flowage rights The right to flood riparian land for groundwater recharge, nutrient replenishment, flood attenuation, *etc*.

flowing well An artesian overflowing well or borehole. It is considered unethical to leave a flowing well uncontrolled, because it lowers the pressure head in the aquifer (possibly over a very wide area), wastes water, and causes hydrocompaction (which is only partly reversible). Some isolated flowing wells in arid areas have become oases, upon which wildlife has become dependent.

flow line = **streamline**. A line orthogonal (perpendicular) to the equi-potential line, representing the direction of flow or potential flow. Somewhat analogous to the 'dip' in solid geology. Water *never* flows *across* a flow line. Flow lines and equipotentials together constitute a 'flow net'.

flownet	A configuration of flowlines and orthogonal equipotentials which may be drawn to illustrate the nature of groundwater flow in either vertical or horizontal projections.
flowstone	Water which is supersaturated with respect to carbonate (*eg.* through temperature, biologic, pCO_2 or kinetic effects) tends to precipitate a smooth carbonate crust of flowstone around the channel section. Caves and spring terraces often display intricate and beautiful flowstone structures. Some authorities draw a distinction between flowstone and rimstone.
fluid potential	≈ Hydraulic potential. The sum of kinetic, potential and pressure energy heads (+ any friction losses) at any point in a flow system.
flume	An artificial channel of known geometry, used to laterally constrict flow, thereby creating supercritical conditions and providing a means of accurately measuring the discharge rate.
fluorescein	(Or Na-fluorescein) = fluorescein di-sodium salt: $C_{20}H_{10}O_5Na_2$. One of the rhodamine family of dyes. A very effective tracer that can be detected at extremely low concentrations, and was therefore useful in both tracing and dilution gauging. However, it is no longer used because it is a suspected carcinogen.
fluoridation	The dosing of fluoride-deficient water supplies with low concentrations of sodium fluoride to provide some resistance to dental caries. Although fluoridation has been practised since the 1930s, there is still some dissent as to whether the fluoridation is justified.
fluoride	The most electro-negative of anions in natural waters. It is normally a minor ion but in rare cases, as

	in some low-Ca volcanic environments in Arizona, concentrations reach major ion status. The highest permissible limit varies with the mean annual air temperature, from 0.5_{warm} to 1.7_{cool} mg.l^{-1} as F$^-$.
fluorosis	A range of health problems caused by the ingestion of excessive F$^-$, mainly from natural drinking water. It varies from mild discolouration of teeth (pink mottling) to skeletal fluorosis (a brittleness of bones), and, in extreme cases, to death.
flush	The boggy part of a hillside which contributes substantial surface and interflow during and just after a rainstorm.
flushed zone	That part of the invaded zone of a recently drilled borehole in which effectively all of the original pore water has been displaced by filtrate.
fluvial	Pertaining to flowing water.
flux	(1) Rate of mass flow, as in groundwater flow or moist air advection. (2) Collective noun for a group of hydrogeologists.
FO	Forward Osmosis. The really clever way to desalinate brackish water, or brine, by making natural diffusion work for you rather than against you. The trick is to make water diffuse across a membrane into a synthetic highly concentrated solution of a volatile salt. This salt can then be driven off and recycled by applying heat, but with only half to a third of the energy required in reverse osmosis. As of 2014 there were still a few 'technical teething problems', but the indications are that FO is likely to be the technology of future choice.

fog	Water droplets in the air small enough to be suspended, and dense enough to reduce horizontal visibility to less than a kilometre. (If visibility is > 1 km, then it is *mist*).
fog drip	Advective interception of fog droplets. This is the only source of water in some parts of the Namibian and Atacama deserts.
foggara	*See* **qanat**.
footprint	The plan envelope which contains a storm's precipitation.
footslope	The low angle slope between steep valley side and active fluvial channel. The term applies to youthful valleys, not to floodplains.
forced convection	Fluid motion caused by mechanical stirring. *See* **free convection**.
ford	= **Irish bridge**. *In wet countries:* where stock or vehicles can cross a river through shallow water. *In arid countries:* a dip in a road to allow passage of a flash flood.
formation	A set of rock strata of spacially consistent lithology (rock type). *See* **lithostratigraphy**.
formation constants	*(Referring to a confined aquifer).* The hydraulic parameters of 'storativity' and 'transmissivity'.
formation factor	A contrast between pore-water and rock-particle properties. It may be defined several different ways,

$$F_f = \frac{\theta^2}{\alpha} \approx \frac{D_w}{D_s} \approx \frac{C_w}{C_s}$$

where F_f is the formation factor, θ is the tortuosity of a porous medium, α is its porosity, D is the diffusion coefficient, C is the electrical conductivity, and the subscripts w and s refer to water and solid particles respectively. The most common connotation is 'bulk electrical resistivity (saturated) ÷ the resistivity of pore water (groundwater)'.

formation water Deep, isolated and more or less connate pore waters; typically saline.

form ratio The ratio of a river's depth to its width.

fossil water Original pore water from the time of rock formation. *cf.* **connate water**.

fountain head *(Obsolete term)* The head in a confined aquifer.

fracking Hydraulic fracturing. An environmentally controversial method of extracting natural gas from productive shales, made possible by the development of horizontal drilling, the capacity to fracture the shale by means of shaped explosive charges, and of injecting the fractures with a gel of porous sand and chemicals. There is much public concern, generally unfounded, that groundwaters will be polluted or depleted by this process. In most cases fracking is undertaken at much greater depths than freshwater aquifers and does not affect water supplies. When undertaken responsibly, fracking is not a serious environmental threat, but careless and/or poorly monitored fracking can result in chemical spills, aquifer salinization and gas leakage.

fractured aquifer A hard-rock aquifer in which groundwater is primarily stored and transmitted through secondary porosity. That is, through joints, faults and miscellaneous fractures. Most fractures are natural, caused by tectonics, compaction, or diagenetic changes in the rock matrix.

Francis turbine   A turbine in which the water is introduced to the runner through a spiral tube (the scroll case) with adjustable guide vanes.

frazil ice   Ice crystals, such as hoar frost, that fall into a supercooled stream or lake act as nuclei for ice growth. This accumulates as spiky aggregates around banks, vegetation and rocks. Frazil ice typically looks like a wet 'slushie.'

FRE   Fractured Rock Environment. A heterogeneous anisotropic aquifer which can only be approximated to a homogeneous aquifer on a large scale.

freeboard   The distance between water level and hydraulic top of a structure, that is the stage rise required for overspill. In designing a dam, retaining wall or other structure, sufficient freeboard, the 'free-board allowance', may be required to allow for wind set-up, breaking waves, *etc.*

free chlorine $Cl°$, $HOCl$, and OCl^- are regarded as 'free chlorine' species, as opposed to combined chlorine, such as the chloramines NH_2Cl, $NHCl_2$, etc. Free chlorine is less stable but more effective as a water disinfectant.

free CO_2   (As distinct from combined CO_2 in the form of HCO_3^- or CO_3^{2-}). Free aqueous CO_2 in contact with limestone may be divided into CO_2 at equilibrium, and therefore unavailable for limestone dissolution, and aggressive CO_2 which will attack limestone.

free convection   = **gravity convection**. Fluid motion, often in the form of paired contra-rotating cells, caused by differential

free energy = **Gibbs free energy**, 'G' in kJ.mol^{-1}. A thermodynamic parameter used to quantify the energetics of a chemical reaction.

free energy of activation ΔG. The energy (heat or light) that must be added to a chemical system in order for a reaction to proceed.

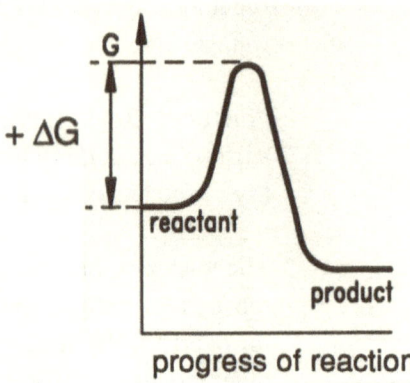

free flow Flow through a structure (weir, flume, spillway, *etc.*) that is not affected by tail-water conditions.

freestone water Water containing little or no dissolved solids.

free surface energy (σ) Numerically and dimensionally equal to the surface tension, 'γ', although they are conceptually rather different. The free surface energy is the work done in isothermally creating a unit area of new surface (as in blowing a soap bubble). The concept has important application in capillarity and soil physics. σ has units of N.m^{-1}.

freezing index On a cumulative 'degree days' *vs* time curve, the annual amplitude is the freezing index. The air freezing index, measured at 1.4 metres above ground,

> surface freezing index (measured just below the ground surface).

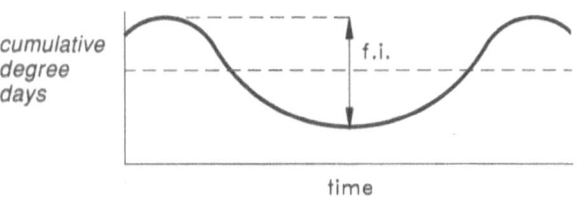

French drain — Also known as a 'French ditch, land drain, rock drain, rubble drain or blind drain'. French drains were originally just a drainage ditch filled with rock, rubble or other porous detritus. More recently the trendy version is essentially a prefabricated tile drain. Some variants are used in urban storm-water or grey-water storage systems. They are also used to drain septic tanks, foundation works or the banks behind retaining walls.

Freundlich isotherm An isotherm of the form $S = K_d.C^n$ (or $S = K_d.C^{1/m}$ according to the convention used), where S is the adsorbed concentration of a solution, K_d is the ion exchange distribution coefficient, C is the concentration in solution, and n typically lies between 0.7 and 1.2. When n = 1, the isotherm reduces to a linear (Henry law) isotherm.

friction factor — Piped network designs are based upon head losses in the pipe, for which engineers tend to prefer the Manning or Hazen-Williams equations, even though they incorporate an implicit false assumption of a constant friction factor. Actually the friction factor varies somewhat with pipe-flow, velocity, pipe diameter, pipe material and presence/absence of a biofilm.

friction slope — S_f = The friction grade, = the slope of the energy grade line. The head loss ÷ horizontal distance of

	flow, in an open or closed channel. In uniform flow, the friction slope = the bed slope.
frost box	A thermally insulating container on / around field instrument-ation to prevent freezing.
frost heave	A huge earth blister caused by local freezing of a groundwater lens.
Froude number	$F_r = \dfrac{v}{\sqrt{g.y}}$, where v is the velocity, g is the gravitational constant (m.s^{-2}), and y is the depth of flow. F_r <1 for subcritical flow, and >1 for supercritical flow. Larger flows have larger critical velocities and critical depths. Criticality indicates whether gravity waves can be propagated upstream or not, and hence whether weirs *etc.* are independent of downstream conditions.
fugacity	'f'. The gaseous equivalent of activity in solution. The fugacity is the effective or thermodynamic partial pressure. This differs slightly from the absolute partial pressure because of the non-ideal behaviour of gases. However, for most water-gas inter-action, the difference between fugacity and partial pressure is insignificant.
full hole drilling	Destructive drilling of the whole bore volume, as opposed to core drilling.
full meander	A continuous full right and left loop of a sinuous river.
fully penetrating	In theory, a borehole which is screened throughout the full thickness of the target aquifer. In practice, any screen which is open to at least the mid 80% of a confined aquifer is regarded as fully penetrating.
fulvic acid	A general class of natural yellowish-brown humic compounds that is soluble under all pH conditions,

but more so under acidic conditions. Fulvic acid is common in groundwaters and surface waters where leaf-litter and other organic débris is microbially degraded, especially in tundra, peat moorlands and forests. Notionally $C_{135}H_{182}O_{95}N_5S_2$, the composition of fulvic acid is actually very variable, with many aromatic and aliphatic components. Nevertheless, fulvic acid has a much smaller molecule than humic acids. It complexes strongly with Fe^{3+}.

fumarole A hydrothermal vent emitting steam and other vapours. Some are associated with a roaring, others with deposition of sulphur.

funicular zone In a granular aquifer the layer between two specific capillary horizons is the funicular zone. The lower horizon is the highest surface at which all the pores are saturated: the upper horizon is marked by the maximum capillary rise (which occurs only in the narrowest continuous pores). In coarse gravels the funicular zone may be only about 1.8 cm thick, whereas in fine silts it may reach 45 cm. The lateral spread of an oil catchment at the water table is mainly restricted to the funicular zone.

GAB The Great Australian Basin is notable as being the world's largest (in area), deepest and most important artesian aquifer, underlying nearly a quarter of Australia. This compares with the worlds' second and third largest artesian aquifers, the Guarani aquifer of southern Brazil and northern Argentina, and the Ogallala aquifer of the American mid-west:

	G.A.B.	Guarana	Ogallala
Area, million km²	1.7	1.2	0.5
Volume in storage m³ x 10¹²	8.7	~30 (?)	5

Groundwater in the GAB drains mainly westwards and southwards, reaching a maximum depth of 3 km. It has an exceptionally low hydraulic gradient, such that the transit time, from the eastern recharge zone to the mound springs in the far southwest, is up to two million years. For about a century the GAB was seriously mismanaged, with declining yields and pressures in uncontrolled boreholes. This is now being remedied, albeit belatedly.

GAB simplified map

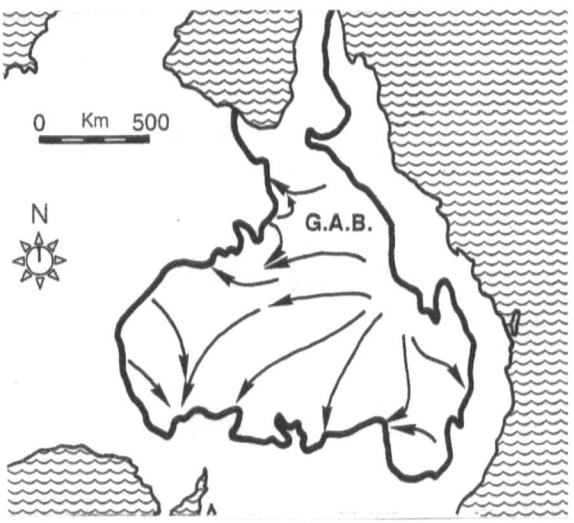

Map of the GAB's perimeter, with groundwater flowlines

GAC — *See* **activated carbon**.

gaining stream — = **Influent stream**. Any reach of a stream or river where there is a nett gain of groundwater to surface flow, as opposed to a losing stream where the reverse occurs. Gaining or losing conditions can be defined

Dictionary of Hydrology and Water Resources

by the configuration of the surrounding piezometric surface.

gallery An adit used for groundwater drainage.

Gallionella A genus of ribbon-like iron-metabolising bacteria, found world-wide in micro-environments from about 1°C to 40°C. It produces ferric iron slimes which foul and clog tanks, pipe-lines, and well screens. It is common in freshwater, seawater and on deep sea sediments. Microscopically it occurs as elongated helical fibres. *Gallionella* thrives best under slightly oxidizing conditions, up to about +300 mV, under cool, hypoxic, pH neutral conditions, where it competes with inorganic oxidation to metabolize iron. Although unsightly, troublesome and difficult to remove, *Gallionella* is not a health hazard.

gallon An obsolete (apart from the USA) unit of fluid volume. 1 Imperial gallon = 4.546 liters, 1 U.S. gallon = 3.785 litres.

galvanic corrosion In addition to normal hydrochemical corrosion, from O_2, H_2S, or various acids, many pumps, rising mains and well-screens have been lost to Galvanic (electrochemical) corrosion. This is caused by the physical contact of two metals with different electrochemical potentials—such as stainless steel and mild steel, or zinc-iron in a damaged GI screen or casing—in contact with an electrolyte (groundwater). The more electronegative of the two metals, the more it behaves as an anode, and the more prone it is to corrosion. The solution is to avoid strong electrochemical contrasts or provide a sacrificial anode.

gamma distributions A family of probability density functions including the Pearson types 3 & 5, log-Pearson types 3 & 5, and generalized gamma 2 to 4 parameter curves. The Pearson type 3 distribution is:

$$f_{(x)} = \frac{(x-x_0)^{\gamma-1}.exp((x_0-x)/\beta)}{\beta^\gamma.\Gamma(\gamma)}$$ where $f_{(x)}$ is the exceedence probability, β and γ are scale and shape parameters respectively, and Γ (γ) is the complete gamma function. Various kinds of gamma distribution are the most widely used pdf's in hydrology.

gamma log The decay of ^{40}K and some other radionuclides involves the emission of very short wave electromagnetic radiation (γ rays). Since potassium is readily adsorbed onto clays, a log of γ distribution in a borehole roughly correlates with clay-rich, low-permeability horizons.

gamma radiation Deeply penetrating, high energy, ultra-short wavelength, electromagnetic radiation. There are only three natural γ emitters of any consequence: ^{238}U, ^{232}Th, and ^{40}K. Their emission rates per gram are similar, but as K is geochemically most abundant, a high γ count in γ borehole logs implies a high clay content.

Gardner's equation This relates the unsaturated hydraulic conductivity to the saturated hydraulic conductivity in terms of a soil factor, A_k, and the capillary pressure,

$$\tau : K_u = \frac{-K_s}{1-A_k\tau^3}(\tau \leq 0).$$

GEF The Global Environment Facility was founded by the UN Development Program, UN Environmental Program and the World Bank. It is a Washington-based international donor whose activities are mainly concerned with biodiversity, climate change and chemical pollution. Many of its funding

Dictionary of Hydrology and Water Resources

beneficiaries are either directly or indirectly related to water. The drawback is that GEF funding is primarily for innovative pilot studies rather than for full implementation of major environmental reforms.

gel Such fine-grained material in suspension that it does not settle without intervention (such as centrifuging, or the addition of a flocculating agent).

gel strength A measure of the ability of a thixotropic drilling fluid to suspend cuttings under static conditions.

geohydrology A needlessly pedantic variant of hydrogeology in which greater emphasis is placed upon surface processes such as infiltration and sediment transport.

geographic information system *See* **GIS**.

geomorphology The science of landforms and their evolution.

geosmin (Trans-1, 10-dimethyl-trans-9-decanol, or $C_{12}H_{22}O$). One of the two most common musty odours from sewage works and polluted waters (the other being MIB), which is produced by the bacterium *Actinomycetes*, and some blue-green algae, such as *Anabaena* and *Oscillatoria*. Although non-toxic, the odour is both foul-smelling and detectable at very low concentrations. In fact about 20% of the population can detect geosmin at a concentration of 13 ppt, and 0.01% of the population can detect it at a concentration of 0.0023 ppt. Consequently, it is often extracted by filtering the water through **GAC** *(q.v.)*.

geostatistics = **kriging**. A technique, usually computerized, to model the spatial variation and magnitude of a parameter, and hence to facilitate the optimum spatial distribution of sampling. This statistical method was developed for mining geologists working with a variable ore grade, but has since been applied to environmental sampling strategies, including optimization of raingauges and borehole networks.

geotechnics The field of soil mechanics and engineering geology, which overlaps with the field of hydrogeology.

geotextile A tough permeable fabric spread over, and pinned to, seepage faces in order to maintain slope stability.

geothermal gradient The rate of change of temperature with depth. The thermal gradient is typically about 25°C per km, but much higher local and regional anomalies occur at tectonic plate margins and especially in geothermal areas. Apart from transient near surface conditions, variation in the geothermal gradient are either caused by heat scavenging in preferred groundwater pathways, or by changes in the thermal conductivity of the rock.

geothermometry Any indirect method (chemical, isotopic or thermodynamic) of estimating the maximum temperature of an underground rock or groundwater system.

geysir (or geyser). An intermittent spout of boiling groundwater found in active hydrothermal systems. The hydrogeology of such systems is complex but requires a water- rather than steam-dominated system and a hydraulic pathway that permits partial cooling, constriction, hydraulic pressure build-up, and cyclic pressure release. The classic examples are found in Iceland, Yellowstone Park (Wyoming, USA), Kamchatka (CIS) and New Zealand.

GGE(e) There are many interactive hydrologic effects related to greenhouse gas emissions. Deforestation contributes up to 30% of the global GGEs, whilst land-use changes both the hydrology and the rates of GHG emission. Paddy irrigation alone causes between 50 and 100 million tonnes of CH_4 emission per year.

ghost bubbles Bubbles of superheated steam that ascend through hot springs in hydrothermal areas, and which condense, collapse and disappear before reaching the surface.

Ghyben-Harzberg relationship The 40:1 ratio for estimating the depth to a saline interface, (unconfined conditions only). It is simply the result of hydrostatic equilibrium of fresh water overlying seawater. The 40:1 ratio does not apply if there is a significantly thick zone of diffusion, giving a salinity gradient of brackish water between fresh and saline. Nevertheless, it works as a first approximation in most coastal and island aquifers.

Giardia intestinalis A water-borne protozoan parasite whose cysts are transmitted by fecal contamination of water, or through cool moist pathways. Giardiasis is not life-threatening, but causes diarrhoea, nausea and debilitation, especially in young children. Giardia is common in Central Australia, in some of the mountain streams of the Eastern Australian highlands, New Zealand, and in cool highland areas of eastern North America.

gigalitre (Gl). 10^9 litres, = a million m³.

GIS **Geographic Information System**. The GIS is an essential part of any modern MIS (management information system) in the water sector. There are countless applications. The basic training can be done in two weeks. Optimum proficiency requires years of experience. Useful though a GIS may be, some applications are still best achieved manually.

For example, this author has yet to see a GIS that plots a realistic isohyetal map. A GIS does not obviate the need for process studies and calibration in the field.

glacial limit = **glacial level**. At any given location the glacial limit is a theoretical altitude above which glaciers could form (assuming sufficient moisture supply).

glacier burst The sudden, and sometimes catastrophic, emptying of an ice- dammed lake. This may be due to either overtopping of the ice threshold, or to lifting of the ice barrier. *See* **jökulhlaup**.

glacier milk Streams fed by glacial meltwater are characteristically heavily charged with white to pale grey, finely abraded rock dust, which gives the water an opaque silvery appearance.

glaciofluvial Generated by, or related to, glacial meltwaters.

glaciokarst Karstic dissolution or fluvial erosion of limestone by subglacial or periglacial streams, and the exposure of endokarst by glacial erosion. Injection of glacial sediment into karstic fissures.

glaciology The study of ice caps, glaciers, icebergs, meltwaters, related geomorphology and associated processes. Glaciology and periglacial hydrology are becoming much more important in the light of climate change. By its nature, field work is inaccessible and expensive, so it tends to be underfunded.

glacis A shallow slope such as an alluvial fan.

glaze = Clear ice, usually formed by the rapid freezing of supercooled water.

gley = Glei. A grey and often mottled glutinous sticky clay soil generated under reducing conditions and poor drainage.

glide A short stretch or reach of smooth flowing water immediately downstream of a pool, in which the surface gradient is positive, whilst the bed gradient is negative.

global climatic model = General Circulation Model, 'GCM'. These computer models consist of an array of some 10^4 to 10^5 grid points, each of which requires multivariate analysis at each time increment. Despite this enormous number-crunching requirement, the grid is still coarse compared to on-the-ground forecasting requirements.

GCM continued Regional subsets of the GCMs have been developed in an attempt to model more local features, giving rise to illusions, amongst the unwary, of 'greater accuracy'. Most GCMs are being continually revised and improved, to take account of emerging feedback processes, better conceptualized atmospheric physics and more reliable calibration. They are highly successful at history-matching broad climatic trends. However, they are designed for global and regional climate change analysis, *not* for water resources analysis—although in practice that is how the latter are often used. GCMs tend to be good at estimating regional temperature changes, but are not so good at estimating trends in rainfall, humidity, and soil moisture.

global radiation The mean solar radiation reaching a horizontal surface at ground level. It is the sum of direct solar radiation (typically about 82%) and diffuse radiation (~18%). Some typical values of global radiation, in $MJ.m^{-2}.d^{-1}$ ($1MJ= 0.278$ KWh), are given below:

	Summer	Winter
Iquitos, Brazil, (Equator)	16.1	16.1
Abu Dhabi, UAE	25.0	14.9
Melbourne, Australia	22.9	6.8
Auckland, NZ	21.3	7.3
Budapest, Hungary	20.1	4.3
Northern Baltic Sea	20.9	1.3
Southern Baltic Sea	20.5	1.6
Scott Base, Antarctica	22.6	0

Note that the global radiation is much less than the solar constant, due to averaging over day and night, atmospheric absorption, oblique incidence, smoke haze, and cloud shadow.

GOES Geostationary Operational Environmental Satellite(s). One of several systems of weather and environmental satellites. There are usually four satellites in operation for remote telemetry and observation.

golden paragraph Career positions in hydrology and water resources have all but disappeared. Most practitioners will need to update their CVs numerous times throughout their career. Assuming adequate experience and qualifications, specialists will be hired—or not—according to their 'golden paragraph' in which he/she describes how perfectly he/she will excel in fulfilling the ToRs from the *employer's* perspective.

gour A rimstone dam. Gours form the cascade of arcuate pools, a metre or two across, which are well known from Rotorua, Yellowstone, and many other mineralised springs around the world. Thermal springs produce the best known gours, but gours are also common karstic features. In fact most are made of calcite rather than silica.

Dictionary of Hydrology and Water Resources

Goyder's Line In 1875 the Australian Surveyor General, George Goyder, published an estimate of the northern limit of grain cultivation. This iconic limit became known as 'Goyder's line', which approximated the 250mm isohyet. Modern equivalents are based upon the ratio of precipitation to evaporation ($P/E \approx 0.26$). In recent decades climate change has been pushing the P/E isopleth southwards (coast-wards), such that about a third of South Australia's grain belt is projected to be lost by 2050.

Gouy layer The outermost layer of the 'electric double layers' which contains a preponderance or + or - ions near an electrically charged surface.

GPR **Ground-penetrating radar.** GPR 'resonates' in air filled cavities, and hence gives some idea of the porosity of dry shallow aquifers. Radar radiation is strongly absorbed by moisture. Therefore GPR works best as a groundwater prospecting tool under hyper-arid conditions.

GPS **Global Positioning System.** A low-cost sensor / computer that obtains positional data from 3 or more satellites. Locations can usually be determined to within about ± 3 metres, or ± about 8 metres in unusual worst cases. Augmented GPS, such as GDGPS, can achieve an accuracy of ± 0.1 metres. Up until the year 2000 the US government had implemented a built-in uncertainty of tens of metres, but this has now been permanently switched off. GPS is particularly useful for gravity surveys, well locations, and general instrumental network systems.

grade The slope of a stream channel surface, bed profile or other ground surface.

Gram's stain A staining technique used on microbial cultures as one of several simple tests for discriminating between bacterial groups. A blue stain is gram-positive, a red stain is gram-negative.

granulometry Pertaining to particle size distribution of a sediment. The granulo-metric curve = the sieve analysis curve. Various particle size ratios such as $U = \dfrac{d_{60}}{d_{10}}, \dfrac{d_{95}}{d_5}$, etc., are taken from the granulo-metric curve, are taken to be characteristic of the sediment, and are used in equations of bed load movement.

graupel Also known as 'soft hail' or 'fluffy snow'. This is a granular cauliflower-like form of accreted snow, intermediate in texture between snow and ice. It is formed by supercooled water droplets freezing onto snowflakes. Individual graupel granules are most typically 50 to 90 μm in diameter but can grow up to about a cm in size. Graupel plays a major role in electrical charge distribution within cumulus convection cells.

gravel pack = Gravel envelope (a little-used term). An annular layer of gravel between a drilled formation and a well-screen. Its purpose is to minimize fine sediment entry into the borehole, whilst allowing the free flow of water through a larger screen slot size than would otherwise be possible. Ideally, the gravel should be rounded, of uniform size, and chemically inert (such as clean quartz).

gravity dam A free-standing massive masonry dam in which the vectorial resultant of the dam's weight, and the horizontal hydrostatic pressure, is contained within the middle third of the base of the dam. That class of dam, as opposed to an earth or thin-shelled masonry dam.

gravity drain	= **field drain**.
greenhouse effect	Without the equilibrium greenhouse effect of the gases CO_2, O_3 and water vapour, the global mean surface temperature would be about -18°C. As of 1990 the global mean surface temperature was +15°C, and is currently on course to reach +19±1°C by 2100 unless we rapidly desist from burning fossil fuels (unlikely). Climate change deniers notwithstanding, the additional anthropic greenhouse effect will drastically alter the hydrology and available water resources of most of the planet.
green snow	A snow surface with a slight greenish colour due to cryophilic green algae.
green wall	Also known as **chroma key**. The blank green background against which a weather forecaster performs, appearing on your TV screens as a detailed weather map. This digital magic is the best way to project a high definition weather map on the same scale as the forecaster. Some 'green walls' are actually blue.
greywater	Water which has been used, but which is still of sufficiently high quality to be of secondary use. For example, apart from sewage, most used domestic water could be recycled for garden irrigation. Greywater contains phosphates and other nutrients that may be a positive advantage for irrigation. Greywater use is likely to increase as the per capita water supply decreases.
grikes	= **kluftkarren**. Joints and other fractures, usually vertical to sub-vertical, in limestone, which have been enlarged by dissolution.

Gringorten formula An expression for estimating the average exceedance probability of a flood, given by: Plotting position (for AEP on a probability scale) $= \dfrac{m-a}{N+1-2\alpha}$ where m is the descending rank order of flood maxima, N is the number of points in an annual series, and α is a constant varying from about 0.40 to 0.45. $\alpha = 0.44$ is adequate for most purposes.

gross duty = **head gate duty** (of water). The total applied irrigation water per unit area, per year or growing season. This includes transmission and channel losses, field waste and evaporation.

gross rainfall The unobstructed precipitation of any given storm or time period.

gross reservoir capacity The storage capacity of a reservoir under normal pool conditions, *including* dead storage, but *not* including surcharge.

ground freezing A method of temporarily sealing a highly permeable saturated matrix to facilitate access for tunnelling, foundation works, *etc*. It is expensive and logistically difficult, but may be the only option for some engineering operations in unconsolidated ground.

ground rainfall The difference between gross rainfall and interception.

ground sill Low walls built across a streambed to inhibit scouring.

groundwater All water held within a saturated soil, rock-medium, fractures or other cavities within the ground. It includes confined and unconfined, saturated and unsaturated, conditions. There are no connotations of residence time, depth or availability.

groundwater artery A preferred groundwater flowpath, usually consisting of a buried coarse-grained fluvial channel.

groundwater divide = Phreatic divide; the groundwater equivalent of a **watershed** (English usage; *q.v.*), but unlike the surface water divide, the groundwater divide can change position. Moreover, the plan trace of the groundwater divide may be significantly different from that of the surface water divide.

grout Any setting or thixotropic fluid. Grouting, an important method for sealing off groundwater flow, involves injection of grout into the matrix. For a sandy matrix the grouting material is usually a cement slurry. For gravels, a swelling clay (usually bentonite) is used. Other grouting materials include special silicates, chrome lignin, and acrylic amides.

grout curtain A series of adjacent wells in which grout is injected to form a continuous impervious barrier. Grout curtains are sometimes used to minimize groundwater seepage away from land-fills or tailing dams.

growing season The maximum number of 'killing' frost-free days that prevails over 80% of years.

growth curve A graph relating increasing relative magnitude (usually of a flood, but occasionally other parameters such as rainfall amount or intensity) to increasing ARIs.

growth curve continued Flood magnitudes are expressed non-dimensionally, such as $\frac{Q_{ARI}}{Q_{mean}}$. Each country or region will have a wide range of different growth curves corresponding to different catchment and climatic conditions. The steepest growth curves are in small arid / semi-arid catchments, whilst the flattest growth curves occur in large wet tropical catchments, or in temperate maritime catchments with consistent weather patterns.

Examples of growth curves:

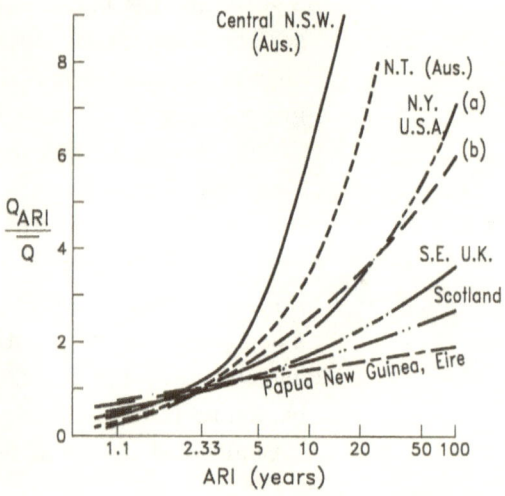

| **Guinea worm** | = *Dranunculus mediensis*. A widespread tropical water-borne parasite whose effects are debilitating and painful. It is contracted by ingestion of unfiltered water containing water fleas. |

| **gulp injection** | 'Single shot' injection of tracer into a stream for dilution gauging, (as opposed to constant-rate injection). Downstream, this results in a bell-shaped distribution of concentration with time (or distance). |

| **Gumbel** | The Gumbel distribution, also known as 'extreme value type 1' or 'EV1' is a much-used two-parameter probability density function. The widely used Gumbel type1 distribution has a fixed skewness of 1.139 which closely approximates the distribution of flood maxima in many environments. Commercial |

	graph papers are available to plot magnitude, or log magnitude, *versus* Gumbel probability distribution.
guttation	If the rate of water supply to leaves, via the root-xylem-air continuum, is > transpiration rate then the stomata fill with water, droplets accumulate on the leaf surface, coalesce and fall to the ground. This process of guttation is most common in tropical rainforests, but also occurs in temperate zone trees, especially in willows. Some fungi and herbs also display guttation.

H

Hadley cell	The planet's largest-scale atmospheric convection system, which advects energy from the tropics to the sub-tropics.

Hagen-Poiseulle eqn. An equation describing the laminar flow of water through a capillary tube, groundwater pore or small pipe. Note the dominant influence of the diameter, 'd'. $Q = \dfrac{\pi d \rho g i}{128 \mu}$ where Q is the discharge rate, ρ is the fluid density, g is gravity, i is the hydraulic gradient (head loss per unit length of flowpath), and μ is the dynamic viscosity. The Hagen-Poiseulle equation is just a variant of Darcy's law that takes account of viscosity in small diameters of flow.

hail (stones)	Precipitation in the form of concentrically layered (clear and opaque) ice balls. Their diameter is usually a few mm, but exceptionally attain about 40 mm.

half life	The time taken for a radioisotope or unstable chemical (such as ozone) to decay to half of its original concentration. *See* 'radioactive decay law'.
Halobacter	An aerobic heterochemotrophic bacterium with a unique photo-synthetic pathway. It requires a saline environment of at least 3 molar Cl^-, and uses a purple pigment in place of chlorophyl. This pigment imparts a characteristic pink to pale purple colour to hypersaline lakes the world over.
halocline	Any layer of a water body which displays a strong vertical salinity gradient. It may, but does not necessarily, coincide with the thermocline. Some very stable lakes have more than one halocline. They are normally thought of as marine, but are also common in saline and brackish lakes. You can make your own visible halocline in a glass half filled with brine, over which you slowly pour fresh water (use the back of a spoon as in Irish coffee). The halocline is the cloudy zone of mixing.
halophyte	Highly salt-tolerant (>0.5% NaCl) vegetation such as saltbush, samphire and mangroves. Halophytes use multiple mechanisms to tolerate salinity, as well as both C3 and C4 photosythesis.
hanging ice	The fusion of multiple icicles to form a ice-curtain or frozen wall of ice. Hanging ice is unstable, so don't attempt to climb it.
hanging valley	A T-junction of a smaller valley truncated against a larger and deeper glacial valley. A common feature in glaciated mountains.
hardness	An archaic and unscientific water quality parameter which, inexplicably, continues to enjoy popular usage. Hardness is the ability of soap to precipitate alkaline

earths (Ca and Mg), and is reported as 'hardness as $CaCO_3$' (in mg.l^{-1}). Total hardness = carbonate hardness + non-carbonate hardness = 50(Ca + Mg) meq. Water with a hardness of < 60 is termed 'soft'. Groundwater from limestone or evaporitic aquifers may have a hardness of 300 mg.l^{-1} or more.

hardware Physical components of a system, as opposed to the 'intellectual input', *e.g.* (1) in computing, (2) in water resources projects.

Hazen unit An obsolete parameter to describe water colour. *See* **colour**.

Hazen-Williams equation An empirical equation for estimating the friction head loss of flow in a pipe: $h_f = \dfrac{10.67 L.Q^{1.85}}{C^{1.85} d^{4.87}}$, the parameters being length and inside diameter in metres, discharge (m^3.s^{-1}), and a constant C, which = 100 for old pipe, and 140 for new iron and cement pipe. (Tables of constants for other materials and conditions are available). Compare the Manning pipe equation. If you prefer a more theoretically-based method, use the 'Darcy-Weisbach' equation.

head Energy per unit weight (of water), measured in units of length.

head-frame The wellhead tackle used for construction of a dug well.

head gate The controlled inlet to an hydraulic conduit.

head-loss The head energy dissipated along a flow path by friction.

head-race The delivery conduit to a turbine.

head-space   The air space between bottled water surface and the bottle's cap.

head-works   All the engineering works required for extraction, storage and transmission (but not including distribution) of water resources. Engineering structures at the inlet to a channel or conduit.

headwater   The upper reaches of a catchment, *or* the water immediately up-stream of a hydraulic structure.

heat island   Urban areas develop a 'heat island effect', most strongly between the surface and about 300 metres. This is caused by anthropic energy dissipation, lower surface albedo compared to surrounding rural areas, higher dust, soot, SO_2, N_2O and water-vapour concentrations. The effect is to increase the urban air buoyancy, and hence enhance convective precipitation. The upshot is that heat islands typically experience more rainfall than the surrounding areas. Local increases of +5 to +20% seem to be quite common, whilst increases of more than 50% have been reported in a few cities. It is not yet clear whether the excess convection of heat islands initiates new convection, or just amplifies convective rainfall that would have happened anyway.

heat of hydration = Heat of wetting, 'H'. Thermal energy exchange involved in either the dissolution of a solid phase, or in simply wetting a solid surface, such as a previously dry soil. Strongly exothermic examples are, NaOH pellets dropped into water, and the hydration of cement. Less dramatically, a wetted clay tightly binds a hydration sheath of water molecules which were previously free to vibrate. Consequently, the kinetic energy of water molecules is converted to sensible heat. The heat of wetting of clays varies from about 5 to 100 $J.g^{-1}$. ΔT (°C) = H ÷ specific heat.

Dictionary of Hydrology and Water Resources

heavy metals The 'heavy metals' include at least: Cd, Cu, Co, Cr, Hg, Mo, Ni, Pb, U and Zn. These are the metal pollutants that pose a significant (sometimes alarming) health hazard in both surface waters and groundwater. Strictly, all the transition elements plus In, Tl and Sn should be regarded as 'heavy metals', although most of this large group of elements occurs at aqueous concentrations far below normal detection limits, even in polluted environments. Be and, to a lesser extent, Al are toxic metals, but are not classed as 'heavy metals'; neither are the non-toxic metals Fe and Mn. On the other hand, As and Se, though not strictly metals, are sometimes grouped with the heavy metals by virtue of their toxicity. Heavy metal pollution has been a major water quality issue since at least Roman times, and continues to be a world-wide source of both chronic and acute toxicity, affecting millions of people.

heavy oxygen ^{18}O. The heaviest of the stable isotope of oxygen, with a natural abundance, relative to the other stable isotopes, of between 1.88×10^{-3} and 2.22×10^{-3}.

heavy precipitation An arbitrary and unnecessary distinction of any precipitation rate in excess of 7.5 mm.hr^{-1}.

heavy water Water consisting of the rare molecule deuterium oxide (D_2O), present in normal water at about 0.02%.

hectopascal HPa. The modern unit of atmospheric pressure, = 1 millibar, 100 N.m^{-2}, 0.001 atmospheres, 10.2 mm H$_2$O or 75 mm of Hg.

heel . . . *of a dam* . . . The upstream basal edge of the structure.

heel ditch A channel constructed
(1) above a slope to prevent fluvial erosion, or

	(2) around a reservoir to intercept inflowing pollutants.
heel karren	Heel-shaped dissolution depressions which form on shallow limestone slopes. With increasing slopes, their development merges with that of rillenkarren.
Hele-Shaw model	An archaic method of modelling two-dimensional ground-water flow by means of a physical analogue, often with liquids of contrasting density.
helicoidal flow	Rotational flow within a river cross-section; for example, due to centrifugal force.
helictite	An irregular or spiral-shaped speleothem.
heliothermal lake	A shallow lake with freshwater overlying brine, in which insolation is mainly absorbed in the brine. The differential salinity needs to be at least 15 g.l^{-1}. Such an interface behaves like a one-way mirror for incoming solar infra-red radiation, with the result that the radiant energy is trapped in the dense lower layer. Some Antarctic heliothermal lakes have extraordinarily large temperature differentials, with the hypolimnion being more than 30°C warmer than the surface water.
Henry law	A linear isotherm: $S = K_d C$ where S is the adsorped concentration of a solution, K_d is the ion exchange distribution coefficient (in this case the 'partition coefficient'), and C is the concentration in solution.
heterogeneity	Having different properties at different points. In reality, all aquifers are heterogeneous, although we all 'assume' homogeneity in order to simplify their analysis.

heterotroph Micro-organisms requiring a simple organic source of carbon as a food source.

higher high water The higher of the two high tides on any given day (similarly lower high water, *etc*).

highest desirable limit (HDL). The preferred maximum chemical concentration in drinking water, as set by such agencies as the WHO, USEPA, EEC or AWRC. The HDL may be set as much for aesthetic reasons as health reasons.

highest permissible limit The legal maximum concentration of a substance in drinking water. It is generally well in excess of the HDL.

Hjulström diagram A much copied diagram of log stream velocity *vs.* log mean particle diameter, which divides into three fields: erosion, transport and sedimentary deposition.

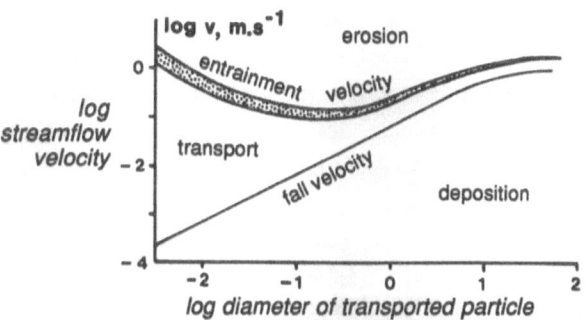

hoar frost Thin ice deposition of distinctly crystalline appearance, white feathery or dendritic deposition on windows.

Holocene The geological epoch since the end of the Pleistocene (ice age: some 10,000 to 18,000 years ago) to the present. Most rivers, lakes and piezometric surfaces at mid-to-high latitudes are still adjusting to post-glacial conditions. Consequently, the Holocene is an epoch of great environmental change.

holokarst Karst development on and in an entirely carbonate formation (as opposed to merokarst).

holomictic Descriptive of any lake which undergoes occasional mixing from top to bottom, (holomixis), in which case it is neither thermally nor salinity stratified.

homogeneity Having the same properties at different locations. A basic assumption of aquifer testing is that the aquifer is homogeneous.

homogeneous halfspace A simplifying geophysical concept in which an electro-magnetic field is centred at ground level. The hemisphere within the hypothetically homogeneous ground is used to describe the depth of sounding and the instrument response. It could be argued that the concept is pointless, because the geophysics is only of value if the ground is non-homogeneous!

hook gauge A means of high precision measurement of water levels, as in an evaporation pan, by means of a metal point raised from below the water surface. As a vernier screw brings the point to the surface, reflected light is distorted by surface tension.

homothermy Temperature invariance with depth.

horizon (1) *In pedology* (the science of soils): One of several distinct layers that tend to develop in a mature soil. At its simplest, the A horizon is the upper zone of leaching, the B horizon is the middle zone of deposition/precipitation, and the deepest C horizon is the layer of weathered bedrock.

horizon (2) *In stratigraphy,* a horizon is an imaginary time line joining points of equal age in a rock sequence.

It need not be horizontal, nor consistent with lithology from one place to another. Hence an upper or lower aquifer boundary may, or may not, be defined by a horizon. In the absence of an abundant diagnostic fossil record, the horizon may be indeterminate.

horizontal well = **trench well** ≈ **water gallery**. A horizontally set screen and gravel pack, usually set just below the water table in a coastal environment. The objective is to skim off the uppermost groundwater with minimal drawdown, thereby minimising the risk of saline upconing. They are particularly appropriate in high-risk thin freshwater bodies, as in coral atolls.

Hortonian flow Overland flow, which occurs when the precipitation rate > infiltration rate, or when the ground is fully saturated.

hot spring *See* **thermal spring.**

Hull formula An empirical estimate of the mixing length (in metres) in turbulent streams and small rivers, required in dilution gauging:
$L = aQ^{0.33}$ (m^3.s^{-1}) where a is 50 for mid-stream injection of tracer, or 200 for 'bankside injection'. *cf.* the 'Rimmar formula'.

humic acid A class of natural organic substances that is characteristically unstable at pH < 2. Humic acid in reservoirs and soil moisture has the ability to form complexes with many of the common cations, especially with Mg, Ca and Fe, as well as with trace metals. Such acids can be very complex with 15 to 20 organic rings as well as sugars and other aliphatic organics. Compare with 'fulvic acid'.

humidity *See* **relative** or **absolute humidity.**

humification Decomposition of organic matter to produce humic substances.

humin Natural organic matter that is insoluble at all pHs.

humus The organic fraction of soil, ($\approx$ humic substances). The humic and fulvic acids in natural waters have chemical similarities to humus, but to assume that natural organic matter in streams and lakes consists of dissolved humus is a gross oversimplification of the natural processes involved. Since it is so stable, humus is sometimes regarded as the end product of organic break-down of microbial and vegetative matter. This is not quite true since the final breakdown products are CO_2 and H_2O. However, humus, especially in the lower undisturbed soil layers, may be essentially unchanged for centuries or even millennia.

hydraulic conductivity (K) A measure of the ease of flow through aquifer material: high K indicates low resistance, or high flow conditions. Units: metres per day. *NB:* K is known in older texts as the 'coefficient of permeability', and is sometimes loosely (and incorrectly) referred to simply as 'permeability'. An obsolete and inappropriate synonym is the 'coefficient of permeability'. *cf.* **permeability**.

hydraulic continuity Hydraulic connection, such as between aquifers, resulting in partial or full hydrostatic equilibrium.

hydraulic diffusivity The ratio of an aquifer transmissivity to storativity, or of hydraulic conductivity to specific storage. This parameter occurs in most equations involving hydraulic response times.

hydraulic fill A method of constructing earth fill dams by transporting the clay core into place by means of flowing water. Following several dam failures the method is now abandoned.

hydraulic fracturing Pressurising a borehole in order to extend natural fractures further into the surrounding aquifer, thus increasing the yield.

hydraulic friction The resistance to flow along a wetted surface of a conduit or channel, but *not* including the head loss due to internal turbulence, sudden obstructions or sharp turns.

hydraulic grade line (HGL). The locus of hydraulic potential along a flow line; for example, the sum of pressure and potential energies along a flow line; a subset of the total energy defined by the Bernoulli equation. It may rise and / or fall along the direction of flow. *cf.* **energy grade line**.

hydraulic gradient In channel flow, the dimensionless mean surface gradient. In unconfined groundwater, the mean water table gradient in the direction of groundwater flow. In confined aquifers, the pressure gradient in the direction of flow—which might not be sub-horizontal. Indeed, it is possible to have a vertical hydraulic gradient.

Hydraulic gradient cont.

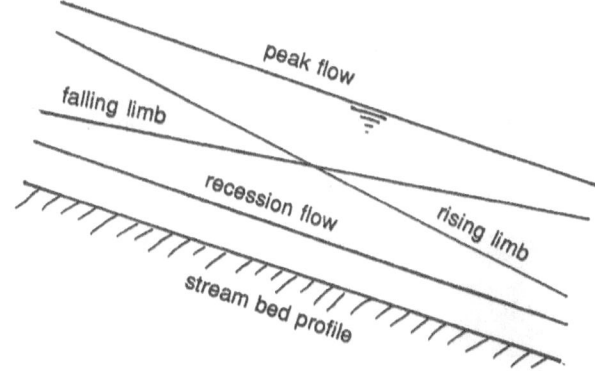

In rivers the hydraulic gradient changes in response to the status of any transient flood wave, as shown above.

hydraulic jetting Backwashing the filter cake in a borehole by means of a high pressure water jet directed through the screen. The objective is to clean up the well and maximise yield.

hydraulic jump An increase in downstream water level (relative to upstream water level) due to the conversion of kinetic energy to potential energy. A form of standing wave. For a constant flow rate, if the upstream depth increases, the downstream depth decreases.

hydraulic mean depth Cross-sectional area of a stream divided by the length of the wetted perimeter.

hydraulic permeability *See* **hydraulic conductivity**.

hydraulic potential Energy per unit weight of a system, expressed in head (metres). *See also* **Bernoulli equation**.

hydraulic radius = **A/P** where A is the stream cross-sectional area, and P is the 'wetted perimeter', used in Manning's, and other hydraulic equations.

hydraulic resistance (c), in units of days. A measure of the resistance of an aquitard to vertical drainage. It is of importance in the analysis of semi-confined aquifer tests, hydro-compaction studies, and in some aquifer models. *See also* 'leakance' and 'leakage factor'.

hydraulic roughness = **Manning's roughness coefficient**, 'n'.

hydraulics The science and technology of flowing water (and, to a minor extent, of other fluids).

hydrochemistry The science and application of water chemistry, including: dissolution, precipitation, adsorption and ion exchange, corrosion, hydrolysis, solute transport, pollution, mass balance studies, phase changes, isotopic systematics, and the behaviour of organic and inorganic solutes. It is a major scientific discipline in its own right. Curiously, despite a vast literature on aquatic solutes and the chemistry of water rock interaction, relatively little has been written on the chemistry of water itself.

hydrocompaction = Consolidation = plastic deformation = discharge per unit cross section area of an aquitard. This process of subsidence is caused by groundwater extraction comprising a combination of water decompression and aquifer compression (mainly instantaneous and reversible), and delayed yield from leaky aquicludes (mainly slow and irreversible). The extent of hydrocompaction varies from mm to several metres. Hydrocompaction is a major contributor to flooding in, for example, Bangkok (cover picture of this book), New Orleans, Mexico City (up to 7 metres in places!), the San Joaquin Valley of California, and several coastal cities of China.

hydrodynamic dispersion A 'bucket term' for the solute transport processes of molecular diffusion, eddy diffusion and dispersion, (as distinct from advection). Dispersion dominates over diffusion, except in very slow moving groundwater. *cf.* **Peclet number**.

hydrogeology The study of groundwater, including flow in aquifers, groundwater resource evaluation, and the chemistry of water-rock interaction. Hydrogeology is arguably the most wide-ranging sub-discipline in the Earth Sciences.

hydrogen ion concentration *See* **pH**.

hydrogen sulphide H_2S, an extremely poisonous and corrosive gas, is occasionally encountered when drilling in strongly reduced aquifers. It is produced both biologically and chemically, but is of only low solubility (<1 mg.l^{-1}), except under highly alkaline conditions.

hydrograph A graph of any hydraulic parameter against time (x axis). Examples include the change of groundwater levels, or the variation of stage, stream discharge or 'log discharge' with time. Hydrograph analysis or prediction is of primary importance in hydrology.

hydrography The measurement of physical oceanographic parameters. Since this is taken to include the tidal reaches of rivers, there is a *very slight* overlap with hydrology. Non-specialists seem frequently to confuse *hydrography* with *hydrology*.

hydrologic(al) cycle A conceptual continuum starting with the evaporation of sea water, moisture advection, precipitation, runoff, groundwater recharge, and finally, surface water and groundwater discharge back to the sea. This is somewhat simplistic, as there are many short circuits, delays and complications, not the least of which is human intervention. Nevertheless, the cyclical notion connotes the essential renewability of most water resources.

hydrological year *see* **water year**.

hydrology The science of water above/on/in the terrestrial environment, and especially the components of the hydrological cycle. In a narrow sense, hydrology is concerned with quantitative time and space-varying surface processes such as precipitation, evaporation, transpiration, soil-moisture storage and fluxes, streamflow, floods, droughts, and catchment responses (natural and human induced). A more general definition would also include: hydro-meteorology, hydrogeology, glaciology, limnology, irrigation technology, hydrochemistry, and water resources assessment and management. In many ways, hydrology is the largest sub-discipline of the Earth sciences.

hydrolysis Dissolution reactions of the form $R\text{-}X_{(dry)} \rightarrow R\text{-}OH + X^- + H^+$ (wet) that occurs when some salts or organic compounds are mixed with water.

hydrometer = **aerometer**. A weighted bulb with an upper graduated stem, calibrated such that the water level of the floating stem indicates the specific gravity of the liquid.

hydrometry The science and practice of water measurement, especially of surface and pipe flow.

hydromorphic Developed under swampy conditions: shallow lakes, marshes *et cetera*.

hydronium The hydrated proton: (H_3O^+). All aqueous hydrogen ions are in fact hydronium ions, but the associated water molecule makes no difference to the hydrogen ion chemistry.

hydrophilic Having an affinity for water; for example, some deliquescent salts, and some micro-organisms.

hydrophilic organics Conservative soluble organic compounds that are difficult to separate from water.

hydrophobic Water-repellent characteristics of solids such as peat soils after a long period of drought, *or* the water-immiscible characteristics of most other non-polar liquids.

hydrophobic organics Organic compounds that readily adsorb onto chroma-tography resins. They are strongly retarded in groundwater flowpaths. Examples include fulvic, humic and tannic acids.

hydrophore A devise for sampling water in a river or lake at a set depth.

hydrophyte Any aquatic plant, especially algae and other immersed biota, but also including plants whose root system is within standing water, or entirely within saturated ground.

hydroponics The practice of growing plants without soil, in which the root zone is suspended in nutrient-rich water, usually supported by coarse sand, gravel, or a synthetic coarse-grained substrate.

hydrosol Any aqueous colloid.

hydrosphere That part of our planet which is actively involved in water circulation, water storage, or water-rock interaction. This comprises the atmosphere up to about 15 km altitude☛, the oceans and Earth surface, and permeable rocks down to a kilometre or two in depth☛. (☛ the exact limits are debatable).

hydrostratigraphic unit Any simple, compound or multiple aquifer mapped as a discrete laterally continuous flowpath. Some hydrostratigraphic units are strongly diachronous, and not a true stratigraphic unit at all. As in chronostratigraphy, hydrostratigraphic units may divide, merge and pinch out, to the confusion of the unwary.

hydrothermal water Literally 'hot water', but there is no general agreement upon what is 'hot'. At one extreme the U.S. definition of *'10°C warmer than ambient near surface groundwater temperatures'* means that some hydrothermal waters still feel cold. At the other extreme, most geologists think of hydrothermal alteration as occurring at several hundred °C. A plausible definition is: water at $\geq$ 35°C, or 'mean annual air temperature + 10°C', whichever is the greater. Water at or above 100°C is chemically very reactive, and involved in many rock reactions that would be insignificant at normal water temperatures.

hyetograph = **hyetogram**. A block diagram of rainfall in successive time units, as used in unit hydrographs.

hygroscopic coefficient The weight % of absorbed moisture in a soil at equilibrium with a relative humidity of 50% at 25°C. It is slightly variable due to hysteretic retention.

hygroscopic water Water held on particle surfaces (by molecular attraction) and which is at equilibrium with atmospheric moisture.

hypersaline There is no consistent definition, but a salinity of >100 g.l^{-1}, or about 30 times sea water salinity, is as good as any. Other authorities define hypersalinity as >10 times that of sea water.

hypolimnetic aeration Or hypolimnetic oxidation; a means of injecting air, and hence oxygen, into the hypolimnion without destabilising the thermal stratification. This is one of several possible remedial measures to improve, or prevent further degradation of, reservoir quality.

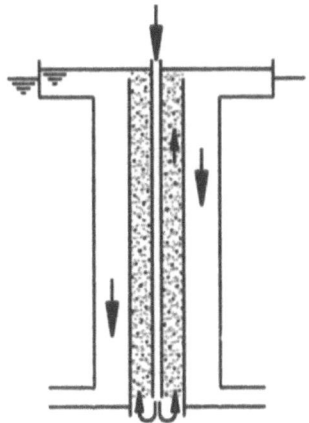

A schematic system of hypolimnetic aeration

hypolimnion That part of a lake which lies below the thermocline. The hypolimnion is typically stable, stagnant, more or less isothermal, and hypoxic if not anoxic. Shallow lakes seldom develop a hypolimnion.

hypoxic Deficiency in, but not total depletion of, dissolved oxygen relative to normal surface waters. Hypoxia is typically caused by the oxidation of organic pollutants.

hypsographic curve A graph of depth (y axis) *vs* % cumulative lake area, as measured from the greatest depth. *i.e.*, at depth = 0, the x axis is n km² or 100%. The hypsograph may be absolute or proportional.

hypsometric curve An *altitude vs.* area graph, as opposed to a hypsographic curve, which is *depth vs.* area. A 'hypsograph' could be either hypsographic or hypsometric. More loosely, a graph of altitude *vs.* distance downstream:

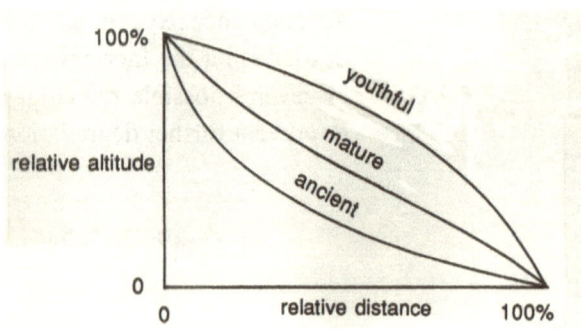

hysteresis	A bimodal graph formed by the soil moisture-suction relationships under wetting and drying conditions. The differential water retention is caused by surface tension effects in irregular pore spaces.

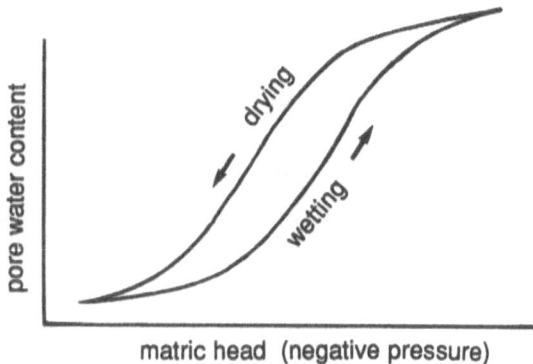

I

ice barrier	= **ice dam**. A partial dam of floating ice blocks, or a total dam caused by glacier flow. Stream flow may bank up behind the barrier, eventually lifting it, and draining rapidly. See **jökulhlaup**.
Ice pipes	Shafts and galleries which form in ice sheets as a result of percolating melt-water, and which are now known to massively accelerate the rate of glacial melt and runoff beneath the ice. Ice pipes, together with refrozen ice, result in anomalous radar reflectivities in Greenland and parts of the Antarctic. In mountain glaciers, ice pipes can grow to cause **jökulhlaups**, (*q.v.*)
ice fall	Part of a glacier under tension, thereby crazing the surface with crevasses.

ice lensing The formation of ice lenses in groundwater on scales from pore volumes to pingos, of > 100 m diameter.

ice pressure The lateral excess pressure against a dam or other structure, caused by the thermal expansion of ice.

ICE WaRM The International Centre of Excellence in Water Resources Management (ICE WaRM). It is an Adelaide-based Australian Government initiative to promote water resources-related training and innovation.

ICM Integrated Catchment Management ≈ TCM: (Total . . .). The coordinated management and sustainable use of all the water, land, vegetation and other resources within a catchment, with particular emphasis on water quality, wetland conservation and storm-water re-use. A popular concept since about 1990.

ideal fluid A hypothetical state of incompressibility and zero viscosity.

igapo A distinctive form of Amazonian rain forest subject to seasonal flooding.

illuviation The precipitation of leachates in a soil's B horizon (= illuvial horizon) from the soil's A horizon, such as Fe, Al and Si minerals.

image well A hypothetical pumped well and associated cone of depression, envisaged as acting opposite a real pumped well, but beyond a recharge or impermeable aquifer boundary. The real drawdown is the sum of the calculated unbounded drawdowns due to the real and image wells.

imbibition The displacement of a relatively non-wetting fluid by a wetting fluid within a granular medium, such as air by oil, or air or oil by water. Plants use imbibition to obtain soil moisture.

Imhoff cone A graduated, conical measuring cylinder of clear plastic and about two litres in capacity. It is used to measure the sediment content of pumped water during well development, or the settleable solids in sewage. It is not a very accurate measure of suspended solids, but it is quick, visual and easy to use.

immiscible = **non-miscible**. The condition where two fluids *cannot mix*, as opposed to the condition where they *do not* mix (such as in density-stratified water bodies).

impellor A rotodynamic machine in which the primary purpose is to impart motion to the water, as opposed to a propellor where the primary purpose is to impart linear motion to the machine. Strictly, the blade of a rotary current meter is neither an impellor nor propellor.

imperfect well A well which only penetrates through part of an aquifer.

impermeable = **impervious**. Not quite such an absolute condition as it might seem, as there is very little in nature which is *totally* impervious.

implicit In numerical modelling an implicit solution is approached by reference to forward as well as backward differences.

import / export Artificial inter-basin transfer of water.

impulse pump A pump, such as a hydraulic ram, which raises water by sudden pressure pulses. The natural pressure regime of a youthful stream can be harnessed to raise small spurts of water high above the normal stream surface.

inactive storage also known as 'inactive capacity'. 'Inactive storage' is *not* the same as 'dead storage'! Inactive storage lies above dead storage and below the active storage. It is there for fisheries or other non-consumptive use, and is lowered only in an emergency or, in the case of small dams, for clearing out accumulated sediment. There may be regulations, governing such factors as the maintenance of minimum hydropower, which legally prevent the use of inactive storage.

in-bank capacity = **bankfull discharge**.

incasion Karstic collapse caused by the continuous cavernous dissolution of limestone aquifers.

incision In regions of tectonic uplift, rivers are rejuvenated and often erode deeply into the river bed, thus restricting lateral movement of the river. Incision may fix a river's course for geologically long intervals.

incrustation = **encrustation**, = **scaling**. A groundwater precipitate, or the process that gives rise to it, which usually consists of iron oxide/ hydroxide slime or crystalline calcite, though many other chemical precipitates are theoretically possible. Millions of dollars are lost every year through incrustation of well screens, pumps, rising mains, headworks, *etc.*

indicator Various strongly coloured chemicals that markedly change colour at a critical point in the mixing of chemicals. The most common indicators are for pH, and are used during titrations. Some examples are:

pH indicator	colour change	pH range
methyl orange (M end point)	red-yellow	3.2-4.4
methyl red	red-yellow	4.8-6.0
p-nitrophenol	colourless-yellow	5.4-6.6
m-nitrophenol	colourless-yellow	6.8-8.6
phenolphthalein (P end point)	pink to colourless	8.2-10.0

Other commonly used indicators are potassium chromate in chloride analysis, phenol red for bromide, and thorin for sulphate. *See also* **mixed indicators**.

indicator organism It is not economically or logistically feasible to analyse waters for all possible pathogens. Instead, initial screening involves analysis of indicator organisms, whose presence implies the probability of enteric pathogens also being present (such as *Vibrio sp.*). No indicator organism is 'foolproof', but in practice, the coliform group in general and *Escherichia coli* in particular, are reliable indicators of faecal contamination. Other indicator organisms include *Pseudomona aeruginosa* and *Clostridium perfringens*.

indirect losses The indirect financial costs of a flood, such as loss of property values, loss of amenity, loss of business, costs of emergency services, *et cetera*.

infiltration The processes of surface water soaking into a soil or rock, or the leakage from a saturated medium to a tile drain or other form of drainage gallery. *See* **limiting infiltration rate**.

infiltration capacity The maximum rate at which a soil or rock surface can absorb recharge water.

infiltration rate The physics of infiltration into a soil is complicated by the soil's heterogeneity, micro-topography, antecedent precipitation and the stage of vegetation.

An initially high infiltration rapidly slows to a quasi-stable rate, which is a function of the rainfall intensity. Graphically, this is of the form:

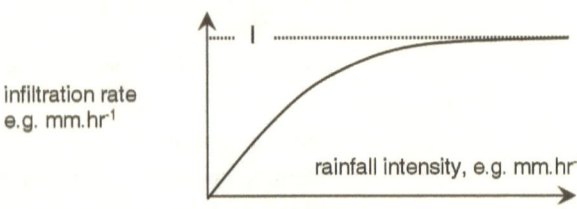

... where I is the maximum *continuing* infiltration rate. *See also* **ring infiltrometer**, and **initial loss**.

inheritance A geomorphologic term for the initial conditions or controls (stratigraphic, tectonic and palaeoclimatic) which influence the development of topography, surface drainage and karstic ground-water flow.

inhibitor Any chemical, added to acid during well acidification, which reduces corrosion of the metallic screen, casing pump, *et cetera*.

initial loss The initial combination of rainfall interception and depression storage which contributes to neither runoff in a catchment, nor to aquifer recharge. It tends to be spatially varied, and hence difficult to quantify satisfactorily.

injection well An artificial recharge well; one that pumps water into the ground.

insequent stream A stream whose direction of flow is erratic; being unrelated to the underlying geology. Insequent stream patterns are dendritic.

insolation The rate of incoming solar radiation received at the Earth's surface; *see* **solar constant**.

insolation ratio The mean ratio of actual to potential sunshine hours $\left(n/N\right)$ for any given place.

instantaneous unit hydrograph The flood hydrograph resulting from 1 mm of effective rainfall falling in an infinitesimal duration. The TUH *(q.v.)* may be derived from the configuration of an IUH.

insular saturation Converse of **pendular saturation** *(q.v.)*

insulosity The % lake area occupied by islands and emergent rocks, sand bars etc.

inter-annual variability A statistical parameter, commonly used to describe variation in the mean annual rainfall. It is defined as: $\dfrac{\sum |X_{i-1} - X_i|}{\mu(n-1)} \cdot 100\%$ where μ is the mean, n is the number of data points, and x_i is the 'ith' value of the data set. *cf.* 'relative variability'.

interbasin transfer With population growth there is an ever-growing need to transport water from where it occurs to where it is most needed, often to distant basins over hundreds of kilometres. Australia, parts of Africa, Canada, China, India, Iraq, and the United States are now very heavily dependent upon interbasin transfer. Where this can occur by gravity it is relatively cheap, but water is heavy and energy intensive, hence expensive to pump, so higher elevation destinations are generally uneconomic.

interception ... or 'interception loss'. The amount of precipitation that falls upon vegetation, and is re-evaporated. For light rain on dense forest, the interception may be 100%.

interception net A medium-to-fine net used to intercept occult precipitation such as sea or moist monsoon mists, or heavy dew. In some hyperarid areas, such low-yielding devices are the only sources of water.

intercropping The practice of planting two crops together; for example shade trees for annuals or tea bushes, or lucerne between date palms. Appropriate intercropping can provide a higher biomass density, fix nitrogen, and make more efficient use of irrigation water.

interference
(1) *In surface water:* Transient changes in the amplitude of a surface wave caused by the intersection of asynchronous surface waves, or of primary and reflected waves.
(2) *In groundwater:* The net additive effects of drawdown and recharge from multiple sources and sinks in an aquifer.

interflow The flow of water along unsaturated flowpaths; infiltration and re-emergence above the water table. Lateral near-surface ground-water drainage that occurs during and shortly after a rainstorm. There are several mechanisms of interflow. In well-vegetated mountainous environments, interflow may account for a substantial fraction of a hillside water balance. Synonyms include: secondary base flow, storm seepage, and subsurface storm flow.

interflux dam A buried dam, (usually) constructed across a fluvial sedimentary section, to control groundwater flow. Interflux dams have considerable potential application in arid regions where ephemeral surface water is limited, and potential evaporation is high.

intensity-frequency-duration A family of rainfall graphs designed to relate these three variables for any given area or region. There is seldom, if ever, sufficient data to plot the most extreme data with high accuracy. Such data is essential for deterministic flood prediction.

intermittent spring An extreme case of an ebb-and-flow spring, in which water appears cyclically, or at least intermittently. The cause is a siphoning mechanism within the groundwater pathway. Such springs generally develop in sub-karstic environments.

internal seiching Resonant oscillation of the thermocline in a lake, in response to wind setup. The hypolimnion, being confined between the thermocline and lake bed, is subjected to greater friction than a surface seiche, and therefore has a longer period of oscillation, given as: $T_i = \dfrac{2L}{\sqrt{\dfrac{gd_h d_e (\rho_h - \rho_e)}{d_e - d_h}}}$ where L is the lake length, the ρ term is the density contrast between hypolimnion and epilimnion, and the d terms are the respective layer depths. In a large lake, an internal seiche may have an amplitude of >10 metres; much greater than that of a surface seiche.

interporosity flow coefficient 'λ' in units of m^{-2}. The interporosity flow coefficient, $\lambda = \dfrac{\alpha r^2 K_m}{K_f}$ where α is a 'shape factor' of matrix blocks in an aquifer, r is the distance to a pumped well, K is the hydraulic conductivity, and the subscripts refer to matrix and fractures. λ is a measure of the ease of groundwater flow into a well which penetrates a double porosity aquifer.

interstices The void spaces in a rock.

interstitial water = pore water.

intrinsic permeability (k). A measure of a soil or rock's capacity to transmit a fluid. Not to be confused with 'K', the hydraulic conductivity. The two are related by the expression

$$k = \frac{K.\mu}{\rho.g}$$

where μ is the dynamic viscosity, ρ is the fluid density, and g is 'gravitational g'. Note that k is a property of the permeable medium, and is independent of the fluid properties (unlike K).

intumescence The swelling of a crystalline substance in response to heating, such as some clays. There is generally an accompanying loss of moisture.

invaded zone An annular volume, around a recently drilled borehole, which is saturated with a mixture of original pore water and filtrate. (The innermost zone is the 'flushed zone').

inversion
(1) A warm air mass overlying a cool air mass, usually with an interface at < 1000 m. This is the reverse of the normal altitude/ temperature relationship between the Earth's surface and the troposphere, and results in much denser air at ground level. The resulting stability creates a kind of atmospheric lid which inhibits the dissipation of urban air pollution and moisture. There are several different causes.
(2) In borehole resistivity logging, thin resistive beds have a curious effect of appearing on the log as a low resistivity, rather than a high resistance, anomaly. It requires a lateral log to correctly interpret such an inversion.

inverted rainfall High altitude rainfall where the precipitation *decreases* with *increasing* altitude (which is inevitable since the temperature, and hence precipitable moisture, also decreases with increasing altitude).

Dictionary of Hydrology and Water Resources

inverted sIphon = sag pipe. A 'U' shaped water conduit used to convey surface flow under roads, etc. Although it has the shape of an inverted siphon, flow is gravitationally driven, and not a siphon mechanism at all.

invisible drought A prolonged period in which precipitation is insufficient to make up the soil-moisture deficit.

ion Any charged atom or molecule. In particular most salts either partially or completely ionise in solution, for example: $KCl \rightarrow K^+ + Cl^-$. Because of the polarised electrical nature of the water molecule, ions attract a hydration sheath of water molecules. However, this does not inhibit the chemical interaction of ions.

ion exchange selectivity coefficient (k_s) is a characteristic constant expressing the ratio of two ions in solution, to their ratio adsorbed onto a surface immersed in, and at equilibrium with, that solution. k_s is a kind of partition coefficient, obtained from: $k_s = \dfrac{mB^+}{mA^+} = \dfrac{m\overline{B}}{m\overline{A}}$, where mX^+ is the molar concentration in solution, and $m\overline{X}$ is the adsorbed concentration.

ionic radii A notional set of radii, based upon crystallographic coordination observations, in which the cations have small radii, and the anions generally larger radii, for example Ca^{2+} 114 and CO_3^{2-} 185pm.

ionic wave gauging A defunct term for dilution gauging.

ion pair A variety of complexed ion in which there is a zero nett charge, for instance: $Ca^{2+} + SO_4^{2-} \rightarrow CaSO_4^\circ$ (aq). After combination, the ion pair continues to be hydrated.

IOS Immobile Oil Saturation, θ_i. The IOS is the fraction of porosity in the unsaturated zone that retains oil as a pellicular film after gravitational drainage from fully oil-saturated conditions; hence, the oil equivalent of field capacity. θ_i varies a little with the type of oil and formation, and with temperature. The typical range of θ_i is from 0.10 for light oils to 0.20 for heavy oils.

irrigation efficiency η. Irrigation efficiency is the ratio of the amount of water a crop actually needs, to the amount of irrigation water actually applied. Note that 'what the crop needs' is more than just the crop water requirement. It also includes a proportion of water to leach out any salt to below the root zone (and thence to appropriate drainage). Most irrigation systems around the world are woefully inefficient—which leaves plenty of scope for water conservation. For well-drained soils and low salinity water, η typically varies between about 0.9 for drip irrigation, and 0.7 to 0.6 for flood irrigation. The efficiency of irrigation tends to vary throughout the annual crop growing season. The concept of irrigation efficiency can apply on any spatial or time-scale.

In addition to common usage, irrigation efficiency can also refer to the response of a crop to irrigation, or to the uniformity of water application.

irrigation excess = **overwatering**. Irrigating at a greater rate than the combined crop water use and reasonable leachate application.

irrigation requirement = **duty of water**. The minimum amount of water required to satisfy the optimum crop water demand, leachate requirement, and any transmission/conveyance losses. It can be calculated on a 'farm gate' basis, or on a total irrigation system basis.

Irish bridge — A dip in a road as it crosses a fluvial channel, built on the assumption of rare flooding. Irish bridges require both up-stream and downstream protection against scouring.

iron — Despite its great abundance in the Earth's crust, iron is generally a minor or even trace element in solution, with a typical concentration of about 10 μg.l^{-1}, and up to tens of mg.l^{-1} occurring under reducing or acidic conditions. Fe^{2+} is the most common highly soluble species, but many complexes exist, both inorganic and organic.

iron-metabolising bacteria Many bacteria obtain their energy by enzymatically or indirectly oxidizng ferrous iron. The unsightly orange-coloured iron slimes, so produced, tend to clog pipes and well-screens, and are generally an economic nuisance, even though the bacteria are non-pathogenic. The most common genera of iron-metabolising bacteria are: *Gallionella, Clonothrix, Crenothrix, Leptothrix* and *Sphaerotilus*.

IRRI — The International Rice Research Institute, at Los Baños near Manila, Philippines, is an independent, not-for-profit research and training institute for the world's most widely grown crop. It is a member of the CGIAR consortium.

ISE — Ion-Selective Electrode/ Electrochemistry. A quick and easy means of direct activity measurement. Halides, and a few metals such as Hg, Pb, Ca, & Na, are the most common applications, but most IS electrodes incur potentially serious interference from other ions. Most modern electrodes consist of a robust plastic housing containing both the reference and sensing electrodes, and a delicate permeable membrane which is open to the solution being analysed.

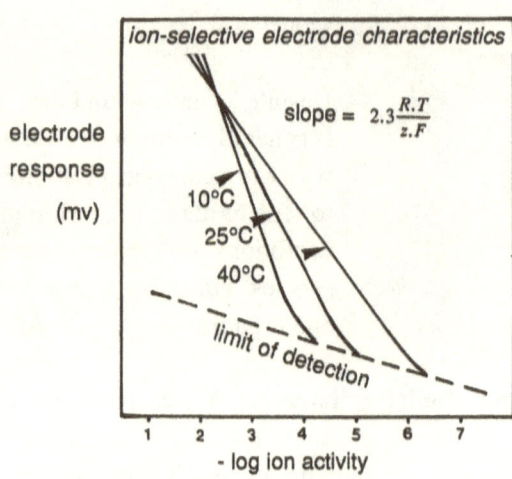

isobath	A line of equal depth, as in the contours of a lake bed.
isocheim	A line of equal winter temperature.
isochion	= **isonival**. A contour of equal water content of snow.
isochlor	A contour of equal chloride concentration.
isochrone	A line of equal travel time of surface flow to some reference point in the drainage system. They tend to form a concentric pattern on a catchment map.
isocon	A contour of equal solute concentration in groundwater.
isohaline	A contour of equal salt concentration.
isohyet	A notional line of equal rainfall. Usually a line connecting points of equal mean annual rainfall.
isonif	A line of equal snow depth.

isopiest A line of equal hydraulic potential, which in unconfined aquifer, is synonymous with a water table contour.

isopleth A general term for a line of equal parameter value on a map.

isothere A line of equal mean summer temperature.

isotherm (1) At equilibrium, the concentration of a substance in solution will differ from the adsorbed concentration (the concentration on the surface of a solid within the solution). The relationship between these two concentrations is described by isotherms, some examples of which are shown in the graph. *cf.* 'Freundlich', 'Henry' and 'Langmuir isotherms'. Consistency of these relation-ships assumes constant pH and constant temperature—hence the term 'isotherms'. Many types of isotherm are possible, as shown below

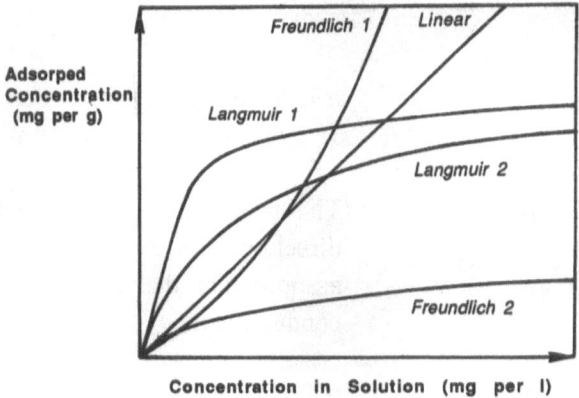

isotherm cont. (2) A line of equal temperature, generally shown as contours on a vertical geological section. Water is such a good scavenger of heat that shallow isotherms are often strongly influenced by groundwater flow.

isotopes Atoms of the same element which have different weights due to slight variations in the number of neutrons in the nucleus.

isotopic exchange At normal temperatures it is doubtful whether any significant isotopic exchange occurs between groundwater and the aquifer matrix. However, under hydrothermal conditions oxygen and carbon isotopic exchange certainly occurs, especially where the rock is a carbonate. Very few water-rock interactions involve appreciable hydrogen isotopic exchange.

isotopic fractionation For any two phases in isotopic equilibrium, such as water-rock or water-vapour, there exists a slight difference in isotopic ratios between the two phases; there is a slight partitioning of a heavy isotope into one or other phase. This is measured by the 'fractionation factor', α.

For example: $\alpha^{18}O = \dfrac{(^{18}O/^{16}O) water}{(^{18}O/^{16}O) vapour} = 1.0115$ at $0°C$

α is not quite a constant, but varies as a function of $1/t^2$, where t is the temperature in °K.

isotropic The property of having equal properties in all directions. For example, most aquifer test solutions assume that the aquifer has isotropic hydraulic conductivity.

isovel A line of equal velocity within a stream channel section.

ISV Ion-Stripping Voltametry. Measurement of trace concentrations of heavy metals in solution. The method concentrates the metals by amalgamation with a mercury-coated cathode. The polarity is then changed, and the various metals 'plated out' from the

Dictionary of Hydrology and Water Resources

mercury anode by metal-specific applied voltages. The concentration step permits a precision to ppb level.

ITCZ Inter-Tropical Convergence Zone. A mean locus of converging lower limbs of northern and southern Hadley cells. The ITCZ is the most important climatic and weather control of the tropics. Its locus follows a seasonal oscillation north and south of the equator.

IWMI International Water Management Institute. Headquartered in Sri Lanka, IWMI is a research organization affiliated with CGIAR, and has 9 regional offices, mostly in Asia and Africa.

IWRM A politically-correct mantra of water management principles which has dominated the water industry, worldwide, since about the 1980s. Integrated Water Resources Management is largely based upon the **Dublin Principles** *(q.v.)*, and involves 'water user participation', 'women's inclusion in water management', and the politically sensitive issue of realistic water pricing.

In practice there are countless river basins, especially in developing countries, where responsibility for water management has devolved from professionals who should know what they were doing, to communities who haven't a clue how to manage their water resource. Many developing country governments have used IWRM as an excuse to reduce their investment in the technical support for the water sector. Originally, IWRM was intended to solve conflicts of over-subscribed water usage, especially in transboundary river basins. In reality there is not a single international river basin in which IWRM has achieved this goal. IWRM has been embraced as essential doctrine by all the major donor organizations (EU, WB, ADB, AusAID, etc), despite its demonstrable failure. It is bizarre that a generation of water professionals have been faced with the choice: 'enthusiastically promote these ineffective principles—or stay unemployed'.

Jacob equation	*See* **Cooper-Jacob equation.**
jars	Interlocked shafts, like two links of an elongated chain, used to jerk free a stuck bit in cable drilling.
jetting	A method of well development in which water is forced through a nozzle (a 'jetting tool') under high pressure. This is often very effective in dislodging the filter cake and other fines behind the well-screen.
jetted well	A well sunk into uncemented sand or silt by sliding casing into a hole as it is excavated by a high pressure jet of air or water.
jökulhlaup	= **glacier burst**. When the water in an ice-dammed lake reaches a critical stage, the ice dam becomes buoyant, and lifts, thus releasing the lake water in a flash flood, or 'jökulhlaup'. The hydrograph of this phenomenon is very distictive in being strongly negatively skewed.
JTU	Jackson turbidity unit. An inaccurate and defunct measure of turbidity, best forgotten. JTUs are superseded by, but numerically approximate, the more precise NTUs.
jump loss	The jump loss, or energy head loss, h_j, across a hydraulic jump, is given by: $h = \dfrac{(y_2 - y_1)}{(4 y_1 y_2)}$ where

y_1 and y_2 are the upstream and downstream depths respectively. The *power* loss across the jump is $h_r.\gamma.Q$, where Q is the discharge, and γ is the specific weight of the water.

junction potential The electrochemical potential generated at (1) the junction of two liquids, (2) water at differing salinities, or (3) two water-saturated lithologies, such as sand and clay.

juvenile water Water supposedly entering the hydrological cycle for the first time. Volcanic emanations are sometimes cited as an example although it is doubtful whether any terrestrial water is truly juvenile. Cosmic hydrogen, oxidised in the upper atmosphere, is probably the only true juvenile water but is totally insignificant volumetrically.

kame A gravel ridge deposited by a sub-glacial or periglacial stream.

kanat *See* **qanat**.

Kaplan turbine A propellor turbine with blades of adjustable pitch to improve efficiency at less than peak load conditions.

Kármán number A dimensionless scaling factor used in models of non-steady stream-flow, to ensure equivalence of roughness between the model and the real case.

$$N_K = \frac{h.v_s}{V}$$, where h is the height of equivalent sand roughness, v_s is the shear velocity, and V is the kinematic viscosity.

Kármán street   The trail of vortices behind an obstruction to fluid flow.

karren   A general term for numerous dissolution features of limestone surfaces. Examples include: humus water grooves, 'heel prints', solution pans, flutes, undercut cavities, grikes, and limestone pinnacles.

karst   A generic term for a huge and distinctive range of geomorphic phenomena caused by rock dissolution in flowing water. Karstic features are mainly associated with carbonate rocks, but evaporites and some siliceous rocks are also karstic. Karstic limestones tend to drain very rapidly, and have medium to exceptionally large transmissivities. Most karstic rocks are strongly anisotropic. Some 10 to 15% of the land surface is susceptible to karstification. *See* **endokarst** and **exokarst**.

katabatic wind   A density current of cold descending air. Such winds tend to be strongly controlled by local relief, and may ruin valley crops in spring or autumn. Examples include winds blowing off ice caps, and cold mountain air descending into valleys at night time. Warm katabatic winds, such as the Chinook and Föhn, may cause unusually rapid snowmelt.

kelly   The square-sectioned uppermost length of a drill string which is free to slide vertically through the rotary table of a drilling rig, and which transmits torque to the 'business end'.

kelly-bushing   In rotary drilling, that part of the rotary table which clamps onto the kelly, and which thereby rotates the drill-string.

Kennedy's critical velocity   The streamflow velocity which will neither deposit nor erode silt from a stream bed. Not to be confused with the more commonly used *stream critical velocity; see* **critical velocity**.

kettle hole A small lake which forms in the depression formed by thawing of an ice lens in periglacial areas.

kemmerer sampler A variety of water-sampling device for retrieving samples from a known depth.

khor Arabic word for drainage channel, watercourse or tidal inlet.

kinematic viscosity v, The dynamic viscosity per unit density of fluid medium. Units of $m^2.s^{-1}$. (= 10 poise in old cgs units). This may seem like a strange and clumsy parameter, but it is convenient in calculations of sediment load, and varying fluid densities.

kinetic energy correction factor 'α' = 'energy correction factor = energy coefficient, = Coriolis coefficient (= 'velocity head coefficient' although this term is best abandoned because of possible confusion with the 'coefficient of approach velocity'). The velocity head of a stream or pipe is given by $h = \dfrac{v^2}{2g}$ or $h = \alpha \dfrac{Q^2}{2gA^2}$ where v is the mean cross-sectional velocity, A is the total cross-sectional area, and Q is the discharge. For deep, slow-moving streams of regular section, α is commonly assumed to be unity. However, α = 1.1 to 1.2 for typical flumes and spillways, is 1.15 to 1.5 for some natural streams, and is 1.5 to 2.0 for streams with overbank flow, conduits with very complex velocity profiles, and mountain streams.

Kjeldahl nitrogen Often referred to as 'total kjeldahl nitrogen' or TKN. The sum of aqueous ammonia and organic-nitrogen N^{3-} species; as opposed to the more common and more oxidised nitrogen species of nitrate and nitrite; *see* **TON**. It is a measure of probable sewage pollution, but is now less commonly used than hitherto.

Klinkenberg effect The slightly higher intrinsic permeability of a matrix to a gas than to water. (A molecular boundary layer effect).

knoll spring *See* **mound spring**.

knowledge products Commonly required deliverables in modern water-sector projects. They may include a hydrologic atlas, explanatory documentaries, a fully populated hydrologic data-base, annual (hydrologic) year book, the maintenance of relevant web-pages, a public interactive centre or a management information system.

kriging *See* **geostatistics**.

kurtosis A shape parameter describing the 'peakiness' of a distribution. The kurtosis, k, is given by:

$$x \frac{n\sum_{i=1}^{n}(x_i - \bar{x})^4}{\left[\sum_{i=1}^{n}(x_i - \bar{x})^2\right]^2}$$

. In the case of a normal distribution, k=3. If k<3, then the distribution is flatter, and if k>3 then the distribution is 'peakier' than a normal distribution.

L

labyrinth A complex flow path which avoids line of sight exposure of an external observer to an internal source of radiation.

labyrinth weir	A complex sharp-crested weir designed to maximise crest length and minimise the fall in head. Even with highly accurate construction, it is of limited precision.

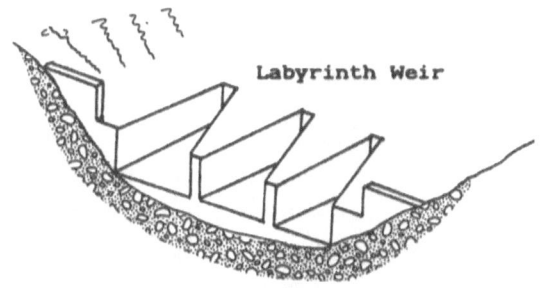

lacustrine	Originating in, or associated with, a lake.
lag time	Delay between peak rainfall and peak runoff.
laminar flow	= Darcian flow, = Viscous flow, = Streamline flow. Orderly, predictable and reproducible fluid flow, with no exchange of momentum, which predominates at low velocities, especially in most cases of groundwater flow. At increasing velocities there is a transition to turbulent flow. With laminar flow, Newton's law of viscosity applies.
landfill	The practice of filling holes in the ground with an avalanche of domestic and industrial waste. This must largely or completely cease within the next century because (1) most of the available sites are already filled, (2) it pollutes our groundwater, and (3) continued production of the current flux of waste is unsustainable.
Langelier Index	An indication of the corrosive *vs* encrusting tendency of a water. It is defined as pH-pHs, where pH is as measured in the field, and pHs is the calculated pH at carbonate saturation. If the index is negative

and there is sufficient dissolved oxygen, then the water would be corrosive. A positive index indicates probable carbonate encrustation.

Langmuir isotherm An isotherm of the form: $S = \dfrac{Q_0 K_d C}{1 + K_d C}$ where S is the adsorbed concentration of a solution, Q_0 is the maximum adsorption capacity (CEC) of the surface, K_d is the ion exchange distribution coefficient and C is the concentration in solution.

Langmuir spiral Paired contra-rotating spiral currents on the scale of a few metres in diameter. They occur on the surface of the sea, medium to large lakes, and sometimes on rivers. They are generated by prolonged strong winds of constant direction. Adjacent spirals result in alternating convergent and divergent surface flow. Convergent flow concentrates foam and detritus into the familiar linear features aligned with the prevailing wind.

Laplace equation $\dfrac{d^2\Psi}{dx^2} + \dfrac{d^2\Psi}{dy^2} + \dfrac{d^2\Psi}{dz^2} = 0$. This describes steady-state 3-dimensional fluid flow in an isotropic homogeneous medium, where Ψ is the hydraulic potential in the axis concerned. The Laplace equation provides a very important basis for groundwater flow analysis and, given well-defined boundary conditions, may be solved either analytically (for simple situations) or numerically. *cf.* **Poisson's equation**.

laser levelling	A 'high-tech' method of accurately levelling land prior to planting in order to maximise the efficiency of flood irrigation. A rotating laser beam is sensed by a receiver on the levelling vehicle, and this raises or lowers a scraper blade on the vehicle to a pre-determined elevation.
latent heat	The energy involved in phase change without any change in temperature. For water vapour the value varies a little with pressure: For values are: water $\Leftrightarrow$ ice 333 kJ.kg^{-1}, and for water $\Leftrightarrow$ steam/water vapour 2250 kJ.kg^{-1}.
lateral log	A form of resistivity log provided by a three-electrode probe. It gives accurate resistivity values, and senses thick and thin beds, but being an asymmetric electrode array, it is relatively difficult to interpret.
laterite	A hard Fe and Al-rich weathering residue typically found in the humid tropics, but also developed elsewhere. Morphologically, it forms nodular, strongly-cemented layers, often closely related to present or past water tables. The most Fe-rich laterites are termed ferrocretes, whilst the most Al-rich laterites are 'bauxites'.
leachate	= Leaching requirement. The irrigation water required to remove accumulated soil salts from the root zone of a crop by dissolution and gravitational drainage to the water table. Expressed in absolute or percentage terms.
leaching	The process of chemical depletion of a matrix by dissolution into infiltrating water.
leaf chloride	Mass ratio of chloride to dry leaf weight (%).

leaf litter Coniferous or deciduous leaf debris, which stores moisture above soils, and which is an important part of nutrient recycling within the biomass.

leaf water potential The negative hydraulic potential (suction), usually measured in the leaf stem, and comprising the sum of the matric and osmotic potentials.

leakance = Leakage coefficient = coefficient of leakage, $=\dfrac{1}{C}=\dfrac{K'}{D'}$ where K' is the vertical hydraulic conductivity of a bounding aquitard, and D' is its thickness (also denoted B or H'). This parameter is used in calculating the hydro-compaction (subsidence) of a depressurised confined aquifer. *cf.* **hydraulic resistance** (c).

leakage factor (B or L) in units of length. The leakage factor, or characteristic length of an aquitard, is an indication of the potential leakage rate into an adjacent aquifer. Large values of L correspond to small leakage rates and vice versa. $L=\sqrt{K.D.c}$ or $B=\sqrt{\dfrac{T.B'}{K'}}$ where T is the aquifer's transmissivity (=$K.D$). By using the leakage factor, the Theis transient equation for drawdown can be modified to take account of leakage: $s=\dfrac{Q}{4\pi T}.W\left(u,\dfrac{r}{B}\right)$

See also **leakance** and **hydraulic resistance**.

left/right bank The bank to the left/right of an observer looking downstream.

legalese insult The despicable legalese practice of doubling up numbers, such as "six (6)". This insults the readers

Dictionary of Hydrology and Water Resources

by implying that they are barely literate. It is a pompous, unnecessary affectation that fails to improve comprehension because it subconsciously causes the eye to backtrack, thereby disrupting the flow of information from page to brain. Alas, it is now universal in all contract documents.

lentic Relating to slow moving or stagnant water such as lakes or swamps (as opposed to lotic). *Latin: lentus* = slow or sluggish.

lethal concentration A parameter used in some bioassay tests of water pollution. The conditions must be strictly specified in terms of species, duration of exposure, and the lethality. For example: *Atrexis spinosa* $_{48hr}LC_{50}$ = 0.3 mg.l^{-1} X, means that on average, 50% of an *Atrexis spinosa* population will die if exposed to substance X, at the given concentration, for a period of two days.

levee A raised river bank. Natural levees have gentle dip slopes (away from the river) caused by fluvial overflow and sedimentary deposition. Artificial levees are built to accelerate flow (reducing siltation) or as flood protection. The breaching of artificial levees during peak floods can be catastrophic.

lift The vertical distance between the DWL and the point of discharge.

LGU 'Local Government Unit', which can be anything from a small community committee to a major city or county council. The client in water projects is frequently an LGU, and is usually very ignorant in water resources issues. LGUs may require some degree of hydrologic or water quality education before they can make sensible decisions.

light water Isotopically light water, free of deuterium. *See also* **heavy water**.

limiting infiltration rate The 'steady state' infiltration rate of a soil after it is wetted to capacity. This is substantially less than the initial infiltration rate.

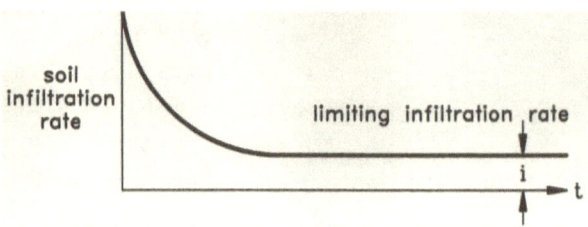

limit of determination The smallest concentration that can be meaningfully quantified in any analytical system. It may be estimated as $x + 6\sigma$, where x is the mean, and σ the standard deviation of the background signal. *See* **lower limit of detection**.

limit of quantitation The smallest concentration that can be quantified with a high degree of accuracy. $m + 10\sigma$. *See* **lower limit of detection**.

limnology *(Greek: limnos* = pool). The study of lakes, and to a lesser extent, of estuaries, springs, and slow moving rivers.

line sink A form of horizontal well in which a porous conduit is placed below the water table in line with the groundwater flow. Drainage may thus be facilitated by tapping the line sink either by gravity (if topography allows) or by pumping from a sump into which the drain flows. Examples are tile drains used in farming, and qanats, widely used in the Middle East for irrigation. A line sink emplaced in, and coaxial with a fluvial sedimentary channel, is the most efficient means of drainage (as opposed to a line sink *across* a channel, as might be supposed).

liquid limit The maximum amount of water that a soil / clay can hold without creeping or initiating flow under standardised 'jarring conditions'. The liquid limit, L_w, is then expressed as the clay's water content as a % of its weight after drying at 110°C. Values vary from about 35% for kaolin to about 700% for Na-smectite. This is one of the Atterburg limits *(q.v.)*

liquidity index $\dfrac{W - I_p}{I_p}$ where W is the natural water content of a soil or clay, and I_p is the Plasticity Index.

liquid junction potential Two fluids of different salinity, such as linked by a porous membrane or pore fluids in an aquifer, set up an electrical potential between the two, which is utilised in S.P. logging.

lithology Rock type.

lithology log = **Litho log** = the combined electric and gamma logs. The screen depths in a borehole are largely decided on the basis of the lithology and driller's logs, although many other logs can also be run if finance and circumstances permit. The hydrogeologist requires experience to make sense of the discrepancies and inconsistencies that arise between the lithology and drillers' logs.

lithostratigraphy A local sequence of rock strata classified by rock type rather than by age. Aquifers are more usefully described by lithostratigraphy than by chrono-stratigraphy, although in some cases, the two descriptive schemes may be essentially the same. The hierarchy of lithostratigraphic units, in decreasing order of magnitude, is: Super-group,

group, formation, member, bed. For example, the Hajar super-group of Eastern Arabia is nearly 3 km thick, whereas individual rock beds may vary between a few metres and a few mm.

litter interception Rainfall intercepted by, and partly re-evaporated from, the leaf litter in a forest or orchard, *et cetera*.

littoral zone This is an ecological term which mainly refers to the marine inter-tidal zone. In the freshwater context it refers to wooded, vegetated and swampy wetlands bordering lakes and rivers.

lockage Water lost to the downstream side of a lock during the course of its normal operations.

loess Highly porous wind-blown silt. In some parts of the world, and most notably in central Asia and northern China, loess has accumulated to hundreds of metres in thicknesses during the Pleistocene epoch. It is easily eroded, and causes most of the world's extremely high suspended sediment loads. The Yellow River alone transports about 1.5 billion tonnes of loess per year, and there are 41 other Chinese rivers each transporting more than a million tonnes of sediment. year[1].

log frame Or 'logical framework', is one of the hallmarks of modern development projects. This is part of 'monitoring and evaluation'; a means of formalizing the obvious. It consists of a 4x4 matrix in which the rows are 'activities, outputs, purpose, and goals', whilst the columns are 'narrative description, objectively verifiable indicators, means of verification, and assumptions'—or variants thereof. Only rarely is this exercise anything more than a time-wasting bureaucratic exercise, but donor organizations like to think that log frames are clever and important. Is it really necessary to discriminate between goals and objectives, *etc*? Tedious though it is, it is necessary to do it in order to win contracts.

log Pearson 'LPIII' Probably the most widely used **p.d.f.** (*q.v.*), and in many countries, the legally required p.d.f. See **gamma distributions**.

long normal A 64-inch probe used in borehole resistivity logging, with good accuracy but poor spatial resolution.

long profile ≈ **Thalweg**. The long profile, the elevation profile of the lowest points of a river bed. Barring geological complications it typically looks like the section below:

losing stream Any stream or reach of a stream in which there is a nett loss of surface water to groundwater. Beneath a losing stream the groundwater isopotentials curve downstream.*cf.* **gaining stream**.

lotic Relating to actively flowing water, such as turbulent rivers, currents or waves.

lower critical velocity The fluid velocity of Reynolds number, $R_N \leq 2320$, below which turbulent flow in a pipe does not occur.

lower limit of detection If any measuring system has a background signal of mean x, and a standard deviation of σ, then the limit of detection is best defined as $x + 3\sigma$. Some authorities use $x + 2\sigma$. If the background noise has a

Gaussian distribution, then these expressions carry a confidence level of 99.98 and 95.44% respectively. Quantitative analysis is not possible below the lower limit of detection.

Lugeon A unit of injection capacity of a formation, comparable to permeability. 1 Lugeon = 1 litre per metre of open hole per minute at 10 HPa pressure. 1 Lugeon ≈ 0.01 m.day^{-1}. A high Lugeon value indicates that a high-density grout will be required for injection. Values typically vary from about 1 to several hundred.

lunette A crescent-shaped sand dune around the downwind side of an ephemeral or seasonal lake, caused by the local redeposition of lacustrine sediments. Gypsiferous lunettes indicate past or current saline lake conditions.

lysimeter One of the oldest forms of field instrumentation in hydrology. A water-tight vessel containing an *in situ* representative 'micro-environment' (such as soil, grassland, crops or even woodland), in which the soil water can be weighed, monitored or sampled. Lysimeters vary from small and simple to huge and sophisticated. In all cases, however, their validity depends upon how well they typify an undisturbed portion of the environment.

maar French dialect for a volcanic crater lake.

macrohydrology The subject of global or very large-scale modelling of atmospheric moisture and rainfall fluxes. Macrohydrology increasingly involves the modelling of climatic change effects, typically on a grid scale of about 400 km x 400 km.

macrophyte A free or rooted water plant, usually of lake or river bank, as distinct from algae.

magnesium Apart from its minor occurrence in water from granitic and arenitic terrains, Mg^{2+} is one of the most abundant ions in natural waters. The range of typical concentrations is about 1 to 100 mg.l^{-1}.

Maha, Yala The two rice growing seasons in Sri Lanka. Yala is the growing season from May to August, when the southwest monsoon occurs. Maha occurs from September to March, during the main northeast monsoon.

main sewer = **trunk sewer**.

major ions The most abundant ions in natural waters, normally consisting of: Ca^{2+}, Mg^{2+}, Na^+, K^+, HCO_3^-, SO_4^{2-}, Cl^-, and (arguably), NO_3^-. Other ions which occasionally attain comparable significance include: OH^-, CO_3^{2-}, $H_2SiO_4^{2-}$, $H_2PO_4^-$, F^-, Sr^{2+} and a range of imprecisely identified organics.

manganese Mn^{2+}. A minor ion in natural waters except under strongly acidic conditions. Its tendency to leave black stains is a nuisance to domestic consumers, and the reason for its maximum permissible limit of 50 mg.l^{-1}. It can be removed by aeration (oxidation), precipitation and filtration. In many respects the aqueous chemistry of manganese is similar to that of iron.

Manning's equation $\bar{v} = \dfrac{1}{n} R^{\frac{2}{3}} . S^{\frac{1}{2}}$ where R is the hydraulic mean radius, S is the hydraulic gradient, $\bar{v}$ is the mean stream velocity, and n is the **Manning's roughness coefficient** *(q.v.)*. Hence discharge $Q = \bar{v}A$, where A is the cross-sectional area of flow. Manning himself didn't like this equation because n is not dimensionless. Nevertheless, the equation works well in practice, and is the most commonly used method of estimating velocity. With modern survey gear a team can generally measure R and S, for a medium sized river, in an afternoon. Be aware that 'n' can vary significantly with stage.

For full-section flow in pipes the Manning's equation can be modified to yield the head loss: $h_f = 6.35 n^2 v^2 L . K_2^2 . D^{-1.33}$ where K_2 is 1.486 in SI units, L is the length and D is the diameter.

manometer An open-ended tube or pipe, usually made of glass, of sufficient inside diameter for capillary effects to be negligible. It may be straight, containing the liquid whose pressure is required, or it may incorporate a U bend containing a dense liquid such as mercury. Pressure fluctuations are reflected by changes in the fluid column. Many manometers may be required to monitor the changes in a complex system.

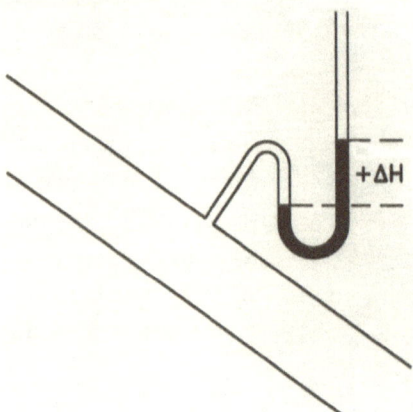

market demand curve The approximate inverse relationship between unit price and the rate of water use. *See* **equilibrium price**.

market supply curve A graph of unit price *versus* total market production rate.

Markov chain Or **Markov process**. A set of data in which each value is, to some extent, dependent upon the preceding value. The data are then said to exhibit 'memory', and can be represented by a 'transition matrix' in which each element is a probability of repetition. Markov chains are critically dependent upon the sampling interval. The weather, stream order, and cyclic lithostratigraphy all display Markov characteristics.

marl A calcareous clay or soft clayey limestone. Marls are often Fe and gypsum-rich, and inter-bedded with shallow marine limestones. Hydraulically, marl behaves as an aquiclude or aquitard, and is often orders of magnitude less permeable than the limestones with which it is inter-bedded.

marsh = **fen**. A seasonally flooded and permanently moist wetland.

Marsh funnel An inverted conical vessel with a hole at the apex, used to estimate the viscosity of a drilling fluid. The time taken for the mud to drain is compared to the time taken for water to drain at the same temperature. The method is imprecise.

formation	*M.f.viscosity (seconds)*
fine sand	35 to 45
medium sand	45 to 55
coarse sand	55 to 65
gravel	65 to 85

masonry dam As opposed to an earth dam. A dam built of stone and / or concrete. There are various types including: thin-shelled dam, gravity-arch dam and buttress dam.

mass balance One of the most fundamental principles in science: Inflow = Outflow ± Change in Storage. For example, this applies to glacier flow, water vapour, sediment load, water flow itself, and to solute fluxes. There are almost endless elaborations on the basic principle.

mass discharge curve A graph of cumulative discharge *versus* time. The mean discharge is given by the mean slope.

mass spectrometer An instrument for measuring the relative abundance of isotopes. Ions of an element or compound are accelerated in a strong magnetic field. Those ions of higher mass (involving the heavier isotope) are deflected less than the lighter ions.

Having split the light and heavy ions into two separate beams, the relative intensity is measured, and the isotopic ratio calculated. The beam radius is given by: $r = \dfrac{1}{H}\sqrt{\dfrac{2V.m}{e}}$ where H is the magnetic field intensity, V is the accelerating voltage, m is the ion mass, and e is the ionic charge.

matric potential A negative pressure potential (suction) caused by capillary and adsorptive forces in the unsaturated zone. This contrasts with the *positive* pressure potential *below* the water table.

mature river Generally, the mid-part of a river's course, where the baseflow is just sufficient for siltation to balance erosion.

Maucha diagram A graphical depiction of major ions in a water analysis.

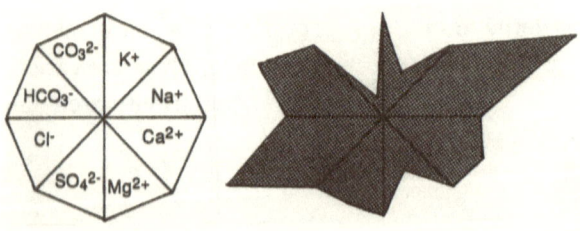

MCA	In the water sector 'multi-criteria analysis' is the application of any of several objective and transparent methodologies, typically used to choose the best alternative amongst various conflicting water uses. The criteria are often grouped under economic, social and environmental, and must each be linked to a quantifiable indicator. Negotiation is required amongst stakeholders to allocate agreed weightings to apply to each criteria. Amongst professionals MCA can become inordinately complex, using non-linear weighting scales, inter-active and time-dependent criteria, *et cetera*. Sophisticated software is available for such analysis. However, most water-users are not sophisticated, and prefer readily understood comparisons. In such cases a simple 'home-made' spread-sheet provides a more than adequate MCA.
MCI	Multiple cropping intensity. The average number of harvests per year, generally 0.4 to 2.1. MCI ≈ 'CHF', the crop harvest frequency, defined as the ratio of annual harvested area to standing cropland areas. Most of the world's MCI>1 is in China, India, SE Asia and central Europe. Much of Western Asia and Southern Africa has an MCI of <0.5.
mean depth	*See* **hydraulic mean depth**.
median	The value which plots midway along a frequency distribution.

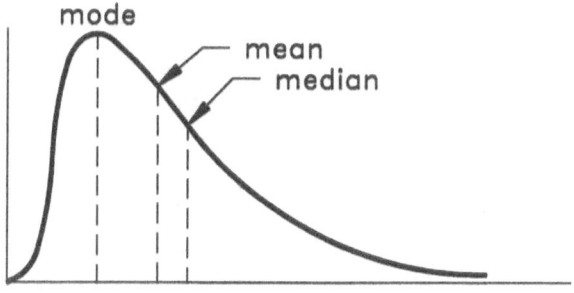

megalitre 1 Ml = 10^3 m^3 (about the volume of a smallish public swimming pool). Probably the most common unit of volume for irrigation and water resources.

meniscus The curved water surface at the water-air junction against a solid surface. Its curvature is most obvious in a vertical narrow glass tube containing a water-air interface.

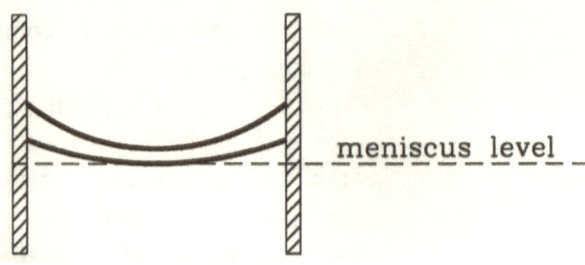

membrane potential The natural electrical potential developed across a clay-permeable rock boundary when both are in hydraulic contact, as in the mud column of a borehole.

merokarst Partial development of karst, associated with layered carbonates and other lithologies. The surface of merokarst contains some features of normal fluvial erosion, together with insoluble residues, and inliers of non-carbonate rocks.

meromictic Descriptive of a lake whose annual (autumnal) overturn only affects the upper part of the lake. The monimolimnion (the deeper part of the lake) remains permanently unmixed.

meromixis The process of mixing and overturn in the upper stratum of a lake.

mesh size The aperture width of holes in a sieve. The mesh sizes (in mm) commercially available are typically: 5.60, 4.75, 4.00, 3.35, 2.80, 2.36, 2.00, 1.70, 1.40,

1.18, 1.00, 0.85, 0.71, 0.60, 0.50, 0.425, 0.355, 0.300, 0.25, 0.212, 0.18, 0.15, 0.125, 0.106, 0.09, 0.075, 0.063, 0.053, 0.045 and 0.038. About eight to ten mesh sizes are sufficient to draw a reasonably accurate graph of the particle size distribution of a sediment.

mesophyte Plants with a water requirement intermediate between the extremes of xerophytes and hydrophytes. This applies to most plants in temperate latitudes and moderately hot and or humid environments.

mesotrophic Intermediate between oligotrophic and eutrophic.

metalimnion A physically important layer of stratified lakes, characterised by the transition from an upper low density layer to a deeper higher density layer. If the density contrast is a thermal effect, then the metalimnion is identified by a *thermocline*. If it is a salinity gradient then it is identified by a *chemocline*.

metamorphosis A 'bucket term' for a range of gravitational and thermal processes which transform fresh low-density snow to a dense snowpack.

meteoric water Water originating from rainfall, usually recent, and hence actively involved in meteoric circulation, as opposed to connate water.

methemoglobinemia Or **infantile methemoglobinemia**. A rare and potentially fatal form of cyanosis to which infants are particularly prone. The main cause is through mothers drinking water with a high nitrate concentration. The nitrate is passed to the baby through the mother's milk, and interferes with the baby's haemoglobin. Thus the baby asphyxiates; hence the common term 'Blue baby syndrome'. The condition can also be caused by antibiotiocs, anaesthetics, and a few other chemicals.

MFI (1) Membrane Filter (or filtration) Index, γ. Originally an index of clogging in reverse osmosis membranes, it has since been applied to the clogging of screens in recharge bores. γ is the slope of the linear part of a graph of $\frac{t}{V}$ vs. V, where t is the time since the onset of water injection, and V is the volume of water injected.

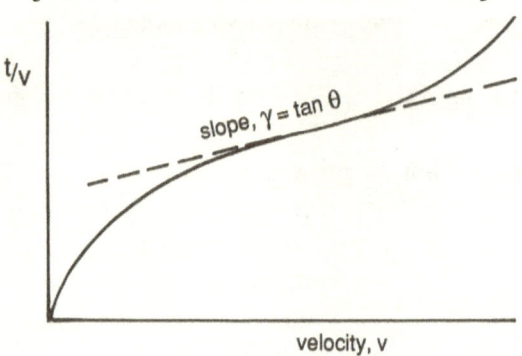

MFI (2) Modified Fouling Index. A parameter closely related to the membrane filter index, but which takes account of temperature, viscosity and pressure: $MFI = \frac{\eta_{20}}{\eta_t} \cdot \frac{\Delta P}{P} \cdot \tan\theta$ where η_t is the viscosity at temperature t, ΔP is the pressure differential across the clogged screen/bore wall, and P is the total pressure.

MIB 2-methylisoborneol, which has the chemical composition of $C_{11}H_{20}O$. An earthy/musty-smelling metabolic by-product of some bacteria and cyanobacteria, with an odour threshold of about 10 ng.l^{-1}. It is one of the two most common foul smells associated with sewage works and organic pollution (the other being geosmin).

MIBK	One of several solvents with a high partition coefficient for transition metal complexes. Hence its use in the APDC-MIBK extraction method for the measurement of trace metals in solution.
micelle	An aggregate of electrostatically-attracted dissolved particles.
microclimate	An imprecisely-defined local climate on a sufficiently small scale to be influenced by measuring instruments, surface turbulence, transpiration from a single plant or wooded area, inhomogeneity of groundwater conditions, *et cetera*.
micro-HEP	Most highland rivers have already harnessed hydro-electric power. However, there is still widespread scope for developing multiple small-scale HEP schemes in high rainfall areas.
micrometeorology	The study of surface and near-surface atmospheric processes, and particularly of vapour and energy fluxes. Micro-meteorology has application in evaporation and transpiration processes. These aspects of hydrology are difficult to quantify, and require expensive and sophisticated instrumentation.
micro-seismicity	Microseismic activity is caused both by subsurface turbulent flow and by fault slippage resulting from increased hydrostatic pressure. The latter is especially prevalent in and around reservoirs where the artificially elevated hydrostatic pressure has been known to cause micro- and macro-seismicity. Micro-seismicity can occasionally be used as a prospecting tool in shallow karstic environments.
mild slope	A channel slope for which a given discharge will flow at a subcritical velocity.

MDGs The UN has promoted eight Millennium Development Goals. None of them explicitly deal with water but, arguably, five of them implicitly involve various aspects of the water sector. It is now clear that the MDGs will not be reached by the target date of 2015, especially in relation to food security, health, sanitation and environmental sustainability—the more so in the light of climate change. In themselves the MDGs are laudable. Alas, that many of the agencies that strive for MDG success are amongst the least efficient bureaucracies on the planet.

milliard 10^9 m^3 = 1 cubic kilometre = 10^6 megalitres. The largest of the commonly used volumetric measures. It is appropriate to use milliards in discussions of regional or global water resources.

milliequivalents 'Milligram equivalents per litre', abbreviated to 'meq.l^{-1}'. An electrical charge concentration contributed by each ion in solution. meq.l^{-1} = concentration in mg per litre x valency ÷ formula weight.

mineral water An extremely misleading term found on bottled drinking water world-wide. 'Mineral water', simply means groundwater, and has no connotations whatsoever regarding the dissolved mineral content. All groundwater contains dissolved salts, and hence, by this legal definition, is mineral water. Surface water also contains dissolved salts but is not mineral water. The joint FAO/WHO food standards programme, which defines 'mineral water' for the water industry, is muddled, fatuous and in drastic need of revision.

minimum response = Minimum speed of response. A low-velocity threshold below which a current meter fails to respond accurately (or at all).

mining An aquifer is being mined if the groundwater extraction rate > mean recharge rate. Formerly mining was regarded as strictly unethical, but mining for temporary water supplies or to increase an aquifer's capacity for ephemeral recharge, may be justifiable.

MIS In modern water management this is commonly used for individual basins, clusters of basins, aquifers or even for whole countries. A good 'Management Information System' is a combination of GIS, water resources database, internet interface, analytical infrastructure (including basin or aquifer computer models), a fast, highly efficient and responsive 'public interface', and most importantly appropriately resourced high-tech water resources expertise to run it all. Even in rich countries, sound water resource MISs are uncommon.

misfit stream A stream that is smaller than one might expect from the size of its eroded river valley. The main causative processes are climate change, river capture and tectonic effects. *cf.* **overfit stream**.

missing data Continuity of data collection is taken for granted in developed countries. Elsewhere, there are invariably gaps in the data caused by lack of instrumental maintenance, loss of instruments, poor training and motivation by/for observers, inadequate water policy (withdrawal of financial support, lack of interest), civil conflict, amongst other causes. Considerable skill and experience is needed to synthesize missing data from adjacent station records, and to judge whether data synthesis is justified, as opposed to leaving blanks in the record. It is *essential* to clearly indicate which parts of the record have been synthesized, and to ensure that the methodology of synthesis is comprehensively documented. It is also necessary to ensure that in any replacement of missing data by synthetic data, the latter is generated with the same amount of statistical scatter as the measured data. Don't even think about synthesizing the missing data until your statistical skills are fairly robust!

mist — A suspension of minute water droplets reducing horizontal visibility, but not as much as with fog.

mixed indicators — The judicious use of two indicators with overlapping colour ranges sometimes yields a more precise end-point than with a single indicator. An example is bromocresol green and methyl red for alkalinity titrations, in which the pH-colour scheme is: light blue 5.0, lavender grey 4.8, pink-grey 4.6, and pink 4.4.

mixing corrosion — If two water sources of differing hydrochemistry, but both at equilibrium with respect to carbonate aquifer material, are mixed, then the resulting hydrochemistry will be undersaturated with respect to the carbonate. Thus some amount of free CO_2 will dissolve carbonate matrix material until a new state of chemical equilibrium is attained.

mixing length — The minimum distance along a river, measured from an upstream injection point, that is required for a tracer to be homogeneously mixed throughout a complete cross section. Naturally, this length varies depending upon the turbulence, and whether the tracer was injected at a bank, or in the centre of the river. Several empirical equations and nomograms are available to estimate the mixing length. None work under all conditions, and all require field verification. For dilution gauging work, it is important to sample at the mixing length (plus a margin for error) from the injection point.

mixing ratio — For a given volume of moist air, the mixing ratio is the mass of water vapour ÷ mass of dry air.

mixolimnion As the name implies, the (upper) mixed layer of a partially mixed lake.

mg.l⁻¹ Milligrams per litre—the standard unit of concentration used in most water analyses. mg.l⁻¹ tends to be synonymous with 'parts per million' by weight (ppm—the unit preferred by some geochemists), though strictly, mg.l⁻¹ and ppm are only numerically equivalent for very dilute solutions at 3.9°C.

mode The most frequent value of a statistical distribution.

modular flow A term used with weirs and flumes: Any discharge in which flow is independent of the tail water levels (= downstream water levels). *NB:* The flow is either 'modular' or 'drowned out'.

modular limit A tailwater condition in which the depth is just sufficient to begin to affect the upstream flow in a weir, flume, spillway, *et cetera*.

moisture equivalent The percentage of water retained in a fine-textured soil after subjecting the soil to centrifugal drainage (of 1000 'G's) for half an hour. As one might expect, there is a rough correspondence between moisture equivalent and wilting point. This term is becoming obsolete.

moisture deficiency Or **field moisture deficiency** is the amount of water required to restore the land in question to its normal field capacity. It is expected that climate-change is likely to increase transient moisture deficiencies in most arable regions, even where the total rainfall is not in decline.

moisture index One of several climatic parameters which indicate the degree of aridity. There are two definitions: (1)

$$MI = \frac{P - ET}{ET} . 100\%$$ where P is the precipitation,

and ET is the potential evapo-transpiration. It may be determined from monthly, annual or mean annual data.

(2) $I = \sum_{i=1}^{12} S_i$, where S_i is the monthly moisture index, defined

as $S_i = \dfrac{0.018 P}{1.045^t}$ where P is the monthly precipitation (in mm), and t is the mean monthly temperature (°C).

molality See **molar solution**.

molar solution A 1M solution of a substance is one 'mole' per litre, which is the molecular weight of a substance, in grams, per litre of solution. In many natural waters it may be more convenient to work in 'millimoles'. *Molarity* is distinct from *molality*, although they may be effectively the same in dilute solutions. The former is weight per unit volume, the latter is weight per unit weight (i.e. g.kg^{-1}).

mollisol A seasonally frozen soil, often highly active.

momentum correction factor $\beta = \dfrac{1}{V^2.A} \int u^2 .dA$ This allows for the velocity of a jet not being uniform over the whole jet's cross section. V is the mean velocity, u is the point velocity. For laminar flow in a tube, β is 1.33. For turbulent flow β is between 1.02 and 1.05.

momentum equation *See* the **Saint Venant equations**.

M&E 'Monitoring and Evaluation' is the politically-correct rolling process of reporting on a project, and determining if it is 'on track'. M&E is invariably a hugely expensive, sometimes futile, and often time-wasting exercise in administrative assessment, which seldom includes scientific assessment. Thus a project can be a *management* success whilst also being a *technical* failure.

monomictic There are two kinds of monomictic lake. In high latitudes, cold monomictic lakes never warm beyond 4°C, and therefore have a cold epilimnion with a single annual mixing event. Warm monomictic lakes occur in subtropical to Mediterranean climates, where the warm epilimnion overturns once a year.

monomolecular film = Monofilm. A thin layer of hydrophobic liquid with good surface spreading properties, such as hexa- or octa-decanol, used to suppress evaporation from reservoirs, and thus conserve water supplies. In small reservoirs under optimum conditions, monofilms can save up to about 25% of the potential evaporation. However, such films don't work on large or windy reservoirs.

monimolimnion Complement of the mixolimnion; the deeper and stable part of a lake in which the upper part has overturned and mixed. *(Greek: monimos = stable).*

monsoon The monsoon is *defined* by the seasonal oceanic-continental air flow, but is usually *recognized* by the monsoon rains. The annual rains caused by moist tropical air drawn into a continent by the development of a seasonal low-pressure zone in mid continent. The classic monsoon region is India, but south-west Arabia, south-east Asia, and north-west Australia also experience monsoon conditions. (from Arabic: *mausim* = season).

monsoon index This has been defined several ways, such as Schick (1953):

$$I = \left(F_{jan} - F_{jul}\right) + \left(F'_{jan} - F'_{jul}\right) \text{ where } F \text{ and } F'$$

are percentage frequencies referring to the direction of the prevailing winds for January and July, respectively. The maximum possible value for this index is 200. A better index would involve weighting of the amount of direction change.

Khromov (1957) proposed an alternative index for those regions where the prevailing wind direction shifts at least 120° between January and July. Then, using average frequencies of the prevailing wind in each of the two months: $I = \frac{1}{2}\left(F_{jan} - F'_{jul}\right)$.

monsoon burst The break of the monsoon season, consisting of hot weather and massive cumulus build-up, followed by the sudden onset of torrential rain.

Moody diagram A family of graphs relating the friction factor, f, to R_e, (Reynolds number), for a range of relative roughness values. It is applicable only to turbulent flow, i.e. where $\log R_e > 3.3$. The relative roughness is the ratio of the surface roughness (in m x10^{-3}) to the pipe diameter. The surface roughness can be determined from tables, e.g. 0.3 to 1 for concrete, 1.5 to 2.5 for rusty cast iron. Hence, the relative roughness and R_e can be calculated. The Moody diagram thus gives the friction factor, from which the head loss in a pipe or channel can be calculated for design purposes.

Moody diagram cont.

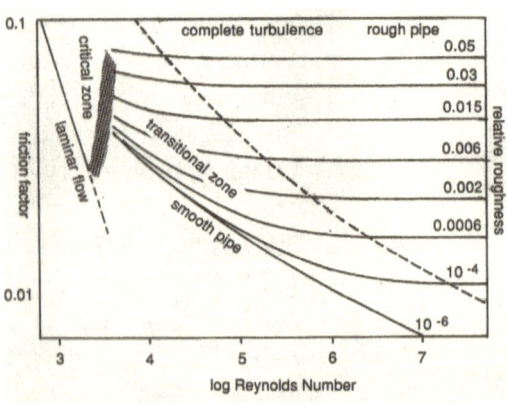

moonmilk	A white aqueous suspension of microcrystalline carbonates (usually calcite, occasionally hydromagnesite or huntite) which forms in caves.
mor	A Norwegian term for coniferous leaf litter or a humus-rich top-soil, largely consisting of raw pine, larch or fir needles, overlying a rock or mineral soil. Mor has a huge ion exchange capacity, and plays an important role in stream acidification. *cf.* **mull**.
mother well	See **qanat**.
mound spring	= **knoll spring**. Accumulations of precipitating minerals by evaporative concentration or cooling from low-discharge springs, usually in arid conditions. Classic examples occur in the western part of the Great Artesian Basin.
moving boat	The only practical technique for measuring the flow of very large rivers. The method requires continuous orientation measurement, flow-metering, and depth monitoring from a boat as it 'crabs' obliquely across the river, whilst compensating for the current.

The moving boat method:

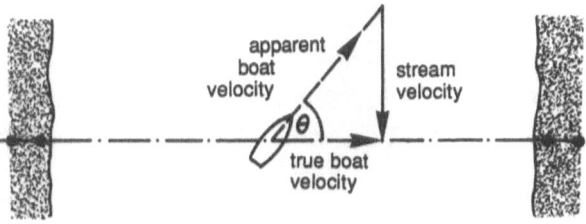

M-point	The colour change from orange to pink when methyl orange indicates a pH of about 4.5, at which point any bicarbonate in solution is assumed to have been neutralised by acid.

MSFL Micro-spherically focussed log. A down-hole inductive-resistivity logging tool with a highly specific horizon resolution.

MSS Multi-spectral scanner. An air or satellite mounted system which scans the ground and records an image digitally rather than photographically. An image is compiled from many thousands of 'pixels', each of which has a radiance recorded at seven wavelengths, varying from IR to UV. MSS systems do not have as good a resolution as photographic systems, but they have the great advantage of facilitating image enhancement, whereby subtle differences in spectral signature can be 'contrast stretched' by subsequent computer processing.

mud pit A small reservoir, dug or transported, for the fluid used in rotary drilling. It doubles as a stilling basin for the rock cuttings.

mulch A porous organic ground cover (such as wood chips or lawn mowings) used to inhibit soil moisture evaporation by
(1) reducing the soil temperature, and
(2) inhibiting the movement of soil moisture into the air.

mull A mixture of rotted leaf litter (especially coniferous) and mineral matter, at moderate pH, and with minor ion exchange capacity; as opposed to **mor** *(q.v.)*

multiple tube An old statistical method of assessing bacteria in water samples. Largely superseded by the plate counting method.

multipurpose reservoir Lakes and reservoirs can be utilized for multiple, and often conflicting purposes, including irrigation, hydropower, urban water supply, flood control, maintenance of environmental flows, fishing, water sports and aesthetics. There is no universal guide to the optimized 'rule-curve' for water level management, whilst any prioritization is likely to drift in response to climate-change. Hydrology meets politics!

muskeg   Hummocky swamp-land (N. American term).

Muskingham method A methodology of flood routing using fluxes and storage, based upon:

$$\frac{I_n + I_{n+1}}{2} + \left(S_n + \frac{O_n}{2}\right) - O_n = S_{n+1} + \frac{O_{n+1}}{2},$$

where I is the inflow, O is outflow, S is storage, and n is a routing period; usually a day. *See* **routing**.

MUSLE   Modified Universal Soil Loss Equation. $Y = \alpha(V.Q_p)^\beta .K.LS.C.P$ This is based upon measured runoff and geomorphic parameters (*cf.* USLE), where Y is the sediment yield (Kg x10^3) from an individual storm, V is the storm runoff in m^3, Q_p is the peak flow rate ($m^3.s^{-1}$), K is a soil erodability factor *(e.g.* 0.12 to 0.18), *LS* is a length slope and steepness factor *(e.g.* 7 to 14), *C* is a cover management factor *(e.g.* 0.21 to 0.18), and *P* is an erosion control practice factor, typically of about 0.65. α and β are empirical constants. Published values include $\alpha = 5.4$ x 10^{-4} to 11.8, and $\beta = 0.56$ to 1.02.

MYTH   An Australian enthusiast group calling themselves by the pompous, somewhat tongue-in-cheek name: 'Meeting of Young Turks in Hydrology'. They correspond by email, and agitate for more rigorous standards in hydrology (badly needed!)

N

Naegleria fowleri — An amoebic parasite found worldwide in some warm stagnant waters. Nasal infection leads to amoebic meningitis (amoebic meningoencephalitis) which is more than 99% fatal, although curiously, it is safe to drink *Naegleria* infected water. The amoeba is controllable by chlorination or chloramination.

nappe (1) The profile of free-falling water over an obstacle, such as a sharp-crested weir.
(2) A geological complex, tectonically emplaced by large-scale low-angle faulting, often involving mountain building, imbrication and rock folding. The tectonics tend to produce complex aquifer geometry, occluded primary porosity, and possibly the develop-ment of secondary porosity.

natural gamma log A borehole profile of natural gamma radiation, as opposed to the γ-γ log which measures back-scattered gamma radiation from a sealed source in the borehole probe.

naturally developed well A well in which the screen is placed directly against the formation with no gravel pack. This is only feasible in a mature lithified or crystalline rock.

natural subirrigation Root systems that extend to the capillary fringe for their entire water requirements.

natural water . . . as described on bottled drinking water, and as ridiculously defined by the FAO/WHO food standards program, means: 'Groundwater which is not derived from a municipal or public water supply'. So if a private well is purchased by a public utility, the pumped water ceases to be natural! Also, by this inane definition, surface water is also 'unnatural'. See **mineral water**.

Navier Stokes equation A fundamental equation describing the flow of viscous incompressible fluid in an ideal, porous medium. In the x direction, it is:

$$\frac{\partial v}{\partial t} = \frac{1}{\rho}\cdot\frac{\partial P}{\partial x} + V\left(\frac{\partial^2 v}{\partial x^2} + \frac{\partial^2 v}{\partial y^2} + \frac{\partial^2 v}{\partial z^2}\right) + F_x \text{ where}$$

V is the kinematic viscosity, and F_x is the driving force per unit mass in the x direction. For some purposes, it is much easier to use the less rigorous (empirical) Darcy equation. Despite the extreme difficulty of analytical solutions, the Navier-Stokes equation has many applications of which the most recent and important is the numerical simulation of mesoscale to large-scale atmospheric and oceanic dynamics. That is, the basis for global climatic modelling.

NBI = Nile Basin Initiative. A 10-nation bureaucracy (technocracy?) whose tag line, or 'shared vision' is *"to achieve sustainable socio-economic development through the equitable utilization of, and benefit from, the common Nile Basin water resources"*. Since 1991 the NBI has undertaken numerous water research and management projects, publishes 'State of the Nile' reports, provides flood forecasts, and acts as the basins' database / MIS. Like all such organizations where the total water resource is over-subscribed, it started with lofty ideals but has been progressively weakened by national self-interest always trumping international cooperation and water sharing.

negative pore pressure = capillary pressure, = tension, = suction.

nest (1) A nest of sieves consists of a stack of interlocking sieves, from coarsest (top) to finest (bottom) **mesh size** *(q.v.)*, which is shaken together to determine a sediment's particle size distribution.

(2) Nested piezometers are a set of narrow bore tappings set at different depths, which are gravel packed and backfilled within a single borehole. The objective may be to monitor variations in vertical head, or to monitor multiple aquifers.

Nett / (net) rainfall = Rainfall excess. Rainfall that actually soaks into the ground. This means gross rainfall - canopy and litter interception - surface evaporation.

neutral stress Stress caused by pore water pressure (as opposed to inter-granular pressure).

neutron log A technique for measuring the hydrogen content (and hence by implication, the water content) throughout a borehole section, by measuring the thermal neutron backscatter from an Am/Pu/Ra-Be emitter. One of several commonly used borehole probes used in borehole geophysics.

neutron meter A field probe for non-destructive and repeatable monitoring of soil moisture profiles. The method involves lowering of a sealed neutron source into previously set observation holes. The neutron backscatter can be calibrated in terms of the soil moisture.

névé Compact or hard granular snow, transitional in texture between fresh snow and ice. Névé forms by diurnal cycles of freezing and thawing, or by pressure metamorphosis of an overlying snowpack. It is characteristic of old snowfields or glaciers.

Newtonian liquid = Newtonian fluid. Most liquids and all gases which conform to Newton's law of viscosity:

$$F = \eta A \frac{\delta v}{\delta x}$$ where η is the coefficient of viscosity, and F is the viscous force, and A is the cross-sectional area perpendicular to the velocity gradient, $\frac{\delta v}{\delta x}$.

Nipher shield A wind-deflection plate, rather like a trumpet bell opening upwards, surrounding a raingauge. It is designed to reduce the aero-dynamic turbulence around the gauge rim.

nitrate The most common aqueous nitrogen species. It is harmless at low concentrations, but at higher concentrations it is implicated in various health problems. Nitrate is also one of the nutrients causing **eutrophication** *(q.v.)* The WHO's highest permissible limit (hpl) for drinking water is 45 mg.l^{-1} as NO$_3$ ($\equiv$ 10 mg.l^{-1} as nitrate-N). The main sources of nitrate in surface and groundwaters are from leached fertilisers, run-off from feed-lots, abattoirs, dairy wastes, acidic precipitation, and from nitrifying cyanobacteria.

nitrite NO$_2^-$. A minor ion in some reduced groundwaters and hypolimnia. In most natural environments, and particularly in turbulent surface water, nitrite is quickly oxidized to nitrate.

nitrogen Up to 15 mg.l^{-1} of N$_2$ gas can dissolve in water at 20 °C, but this form of nitrogen is inert except under unusually bio-active environments. Free nitrogen isn't included in normal water analyses, but other nitrogen species, mainly nitrite, nitrate, ammonia and proteins, are important. Nitrogen species are introduced to water from sewage treatment, farm effluent, municipal and industrial wastewater, and from vehicle exhausts.

niveau A terrace level, usually in the context of a stepped, incised or residual terrace amongst several terrace levels.

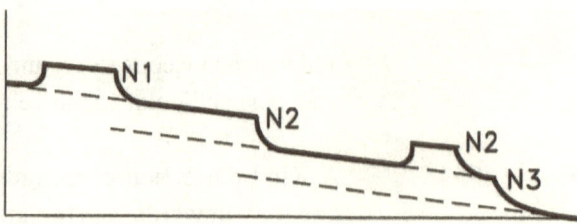

noble gas The inert gases He, Ar, Ne, Xe, Kr and Ra. Their absolute and relative solubilities are sometimes a useful tool in hydrochemical studies.

nodal point In some fluvial systems a transition occurs between the highlands, where streams converge, and alluvial flats where braided divergence occurs. Any clearly defined transition is a nodal point.

node (1) A point of zero or minimum water level fluctuation, as in a seiching lake.
(2) Any point in a spatial array. The properties at that point are taken as representative of the properties over an area. Most 1 to 3-dimensional computer models utilize an array of nodes with specified properties.

Nodularia A common genus of toxic or potentially toxic cyanobacteria, sometimes responsible for algal blooms.

non-carbonate hardness = Permanent hardness. A pointless parameter introduced to account for the common discrepancy between total hardness and carbonate hardness.

non-contributing area Part of a catchment which contributes neither surface runoff nor groundwater flow. Examples include areas of intense rain-shadow, low hot sub-catchments, or areas of internal drainage.

normal depth The depth at which uniform flow occurs, hence the depth at which a stream surface and stream bed are horizontal (assuming constant section dimensions).

normal pool level The normal operating elevation of a reservoir surface. The *minimum* pool level is the minimum level at which a reservoir is allowed to operate. This may be for water quality, slope stability or for pragmatic management reasons.

normal solution The 'normality' is the number of *equivalents* of solute per litre of solution. For example, a 0.5N HCl solution = 0.5 M HCl solution, whereas a 0.5N H_2SO_4 solution = a 1M H_2SO_4 solution.

NOx An undefined blend of nitrous and nitric oxides which oxidizes and hydrates in the atmosphere, and combines with SO_2 to generate acid rain. Both natural and human induced emissions (from burning fossil fuels), are of similar magnitude, at about 50 million tonnes globally per year. However, most N_2O is generated by microbial action in cultivated soils. The rate of N_2O emissions has increased by 74% (1979 to 2013) as a result of increasing fertilizer usage. The greenhouse warming potential of N_2O is 310 times that of CO_2.

NTU Nephelometric Turbidity Unit. The standard comparative measure of turbidity (cloudiness) of water. ≈ **Jackson Turbidity Unit**.

nucleation The formation of raindrops around dust nuclei.

nutrients The plant and bacterial nutrients are nitrogen (mainly as nitrate), phosphate and potassium. All waters contain these in small or trace quantities; at higher concentrations they become troublesome.

oasis A water hole in a desert caused by a topographic depression intersecting the water table, or by an artesian overflowing borehole.

obligate aerobe A micro-organism that must have free oxygen to live.

obsequent stream A stream that flows in the opposite direction to the stratal dip slope, such as down a scarp face. Obsequent streams tend to capture the headwaters of consequent streams.

observation well A narrow-bore well or piezometer whose sole function is to permit water level measurements. In the case of wellfield management it may be the dynamic water levels that require monitoring, but more often observation wells are required to monitor the static water levels over a large aquifer area. In the latter case it is important to site observation wells beyond the radius of influence of any pumped well. Conventionally observation wells are screened to about the mid depth of an aquifer. *See also* **nested piezometers**.

occluded front A meteoric frontal system in which the cold and warm fronts have merged, resulting in one rather than two rainy periods.

occluded porosity Erstwhile porosity which has subsequently been 'filled in' by natural cements, such as calcite, quartz, clay, or iron minerals.

Dictionary of Hydrology and Water Resources

occlusion The process of silting up in a reservoir. All reservoirs act as sediment traps, and hence their expected economic lifetime is closely related to the rate of occlusion. Occlusion rates are notoriously difficult to quantify accurately. Numerous theoretical and empirical equations are available. Most of the latter fail because their calibration doesn't take account of peak sediment transport, which may average only a few hours per year. The only reliable method is that of successive precise bathymetric surveys of the reservoir, before and after a representative number of years.

occult precipitation Water droplets sufficiently small to be transported laterally by advection rather than falling as raindrops (*cf.* **Stokes' Law**). This water re-evaporates or is intercepted by vegetation, and is thus largely or entirely missed by conventional rain gauges, accounting for some of the 'undercatch'. In extreme cases this may be up to 25% of the total precipitation. *cf.* **Scotch Mist**.

offset Wenner A geophysical technique for resistivity surveying, in which two apparent resistivity values are measured with the Wenner electrode array of the second measurement offset by one inter-electrode distance ('a'). The averaged ρ_a for point X largely eliminates the effects of near surface local anomalies, such as the high resistance of boulders.

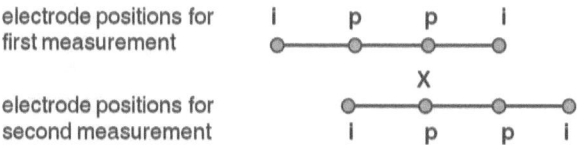

ogee A smooth 'S shaped' double curve, characteristic of spillways and some special purpose sharp-crested weir profiles. It is difficult to make and deploy accurately, but has the advantage of good precision and sensitivity under both low and high discharge conditions.

oil field brines Formation groundwaters associated with, and possibly co-evolved with, hydrocarbons. Such brines are usually hot, strongly reduced, hypersaline and enriched in minor and trace elements. The chemistry of oil field brines is markedly different from sea water or salt-lake brines.

oil retention capacity U_0. A measure of oil pollution in a porous matrix: the oil equivalent of 'field capacity'. Although the orc is slightly time-dependent, it may be taken as the one month residual saturation content.

U_0 has units of litres.m^{-3}, that is of 10^{-3}. Naturally, highly volatile oils, of lower viscosity, have lower orcs. Typical values are 10- 50.

old river A river which has reached the final stages of its geomorphic evolution, and meanders within a broad flood plain in which sedimentation predominates over deposition. The expression has nothing to do with the river's actual age!

oligomictic Descriptive of a tropical deep lake with a stable warm epilimnion. Thermal instability and mixing rarely, if ever, occurs.

oligotrophic *(Greek: oligos* = few, *trophos* = nutrition) Referring to a lake or river water with low nutrient concentrations. Oligotrophic waters are typically clear, fresh, deep, well oxygenated, with low biomass but high biological diversity, such as clear mountain streams. *cf.* **eutrophic**.

open ended well A well / borehole in which the lower end of the casing is left open to the natural formation for ease of entry of groundwater. Obviously, this is only feasible in consolidated formations.

optical raingauge A precipitation gauge that correlates the attenuation of an infra-red beam with the precipitation rate. It has the advantage of avoiding the problem of aerodynamic undercatch, but is expensive to install, and requires a continuous power supply.

optimum shape ... *of open channel flow* ... The channel section carrying the greatest discharge for the smallest cross-sectional area. Other factors excluded, this may be the cheapest channel section to construct. For sloping banks it is half a regular hexagon, whilst for a rectangular channel, it is depth = half the width.

organic carbon Usually 'total organic carbon', or TOC. Aquatic carbon, other than CO_2 species, which usually occurs in the form of humic and fulvic acids, or as organic pollutants.

organic nitrogen A potentially misleading term: all N^{3+} forms of nitrogen in aqueous organic compounds.

organochlorides Also known as organochlorines. A class of insecticides, many of which are transported and bioconcentrated in the aquatic environment. Consequently, many of them are now banned. Examples include DDT, which is suspected of linkage to Alzheimer's disease, and benzene Hexachloride (BHC). Organo-chlorines have characteristically low solubility in water but a high octanol partition coefficient, 'K_{ow}'. That is, a tendency to concentrate in lipids. It may take years or decades for organo-chlorines to work their way through the terrestrial ecosystem to the marine ecosystem.

Organophosphates A group of about 50 phosphate esters which deactivate the essential enzyme acetylcholinesterase particularly in insects. Hence, in addition to their use as nerve agents, organo-phosphates have become 'big business insecticides'. Unfortunately, they are also highly damaging to the human nervous system. In farming and agriculture the choice of insecticide is largely between highly toxic organophosphates, which break down rapidly, or low toxicity (to humans) but environmentally persistent organochlorines. Organophosphate residues in irrigation water have become a major environmental issue.

organoleptic   Pertaining to the senses of taste or smell.

orifice bucket   A cylindrical reservoir receiving a water supply, such as the discharge from an aquifer test, and containing several outlets at different levels which can be screwed open or closed. Tables can convert the internal water level, associated with any combination of outflows, to the discharge rate. It is a primitive method of flow measurement, but has the advantage of damping out fluctuations from a pulsed inflow.

orifice plate   A flow gauging method which utilises the pressure differential either side of a pipe obstruction in the form of a plate with an orifice of lesser diameter than the pipe.

organotin   A range of toxic organic tin compounds of which the most serious pollutant is tributyl tin, 'TBT'. Since the main use of TBT is in anti-fouling paints on ships' hulls, most organotin pollution problems are found in marine environments. However, TBT is also known to be a highly toxic 'endocrine disruptor' and immuno-suppressive in lakes and rivers, with well documented impacts upon fish and many other aquatic organisms. TBT is readily absorbed in lipids, and binds strongly to sediments, and hence is regarded as a persistent pollutant. It has a low solubility in water.

orography High relief causes airflow to rise, cool, and precipitate more than over low relief areas. Most of the world's high-precipitation areas are, at least in part, caused by orography.

orthophosphate (PO_4^{3-}) The final and most stable dissociation product of phosphoric acid, H_3PO_4. Other phosphate species are only converted to orthophosphate above about pH11, but many analysts find it convenient to report the various phosphate species as equivalent amounts of orthophosphate.

osmosis Given two solutions of high and low solute concentrations, separated by a semi-permeable membrane, osmosis is the process of solute diffusion from the higher to the lower solution.

osmotic adjustment The uptake of solutes in a plant to compensate for transpiration stress.

osmotic potential = Solute potential. The negative hydraulic potential caused by increasing the solute concentration.

osmotic pressure = **osmotic potential**. The molecular pressure of a solvent which drives osmosis; *see* **osmosis**. It is measured as the pressure applied to the 'low solute fluid' in order to prevent diffusion from the fluid of higher solute concentration.

outlier (1) A statistical outlier is an extreme value that may, or may not, conform to the same probability distribution as the rest of the data. How to deal with outliers is a major problem in flood hydrology.

(2) A geological outlier is an isolated area of young strata overlying older strata. Porous synclinal outliers are sometimes valuable aquifers.

outwash The fluvial sediment accumulated in a piedmont or periglacial area as a result of a reduced gradient, and hence reduced sediment-carrying capacity.

overbank flow = **out of bank flow**. The portion of a flood that flows beyond and sub-parallel to the main channel, but within the bounding flood plain. Streamflow in excess of bankfull conditions.

overdeepening An erosion feature of glaciers, in which the upper part of a glacial valley is eroded by glacial scouring and plucking, to a greater depth than more 'downstream' parts of the valley.

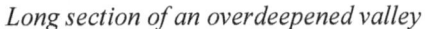
Long section of an overdeepened valley

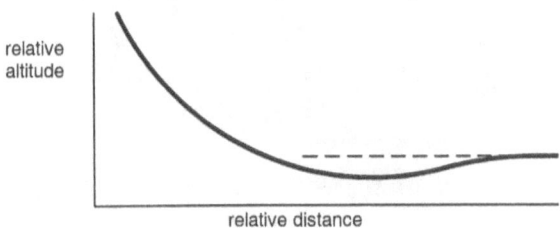

overdevelopment The condition of a well or wellfield in which excessive drawdown renders the yield uneconomic. Local overdevelopment may occur even though the aquifer as a whole is well within its sustainable yield. This hazard is particularly associated with low transmissivities or interfering cones of depression.

overirrigation Application of irrigation water in excess of the crop water and leachate requirements is not just a waste of water, but also a waste of leached humus and nutrients.

overland flow	= **sheet flood**. This is a rapid response to rainfall in the form of sheet (as opposed to channel) runoff. Overland and channel flow behave quite differently; the former being shallow, relatively slow (low momentum), and impeded by high friction. Overland flow usually ceases after a few hours but may last for months in the case of large basins with low gradients, such as the Sudd swamps of the Nile, or the Lake Eyre basin of Australia. There may be no clear distinction between overbank and overland flow.
overfit stream	A river which seems too big for the valley in which it flows.
overpumping	(1) Well development by pumping at a greater rate than the subsequent normal pumping rate. (2) Prolonged pumping at a rate greater than the long-term sustainable rate; 'mining'.
overturn	If / when the density contrast between epilimnion and hypolimnion becomes sufficiently small, it takes only a little wind shear to create partial or total mixing - the 'overturn', of a lake. This is a most important feature of lake dynamics.
overyear storage	A reservoir storage volume capable of meeting a year's continuous demand without replenishment.
oxbow lake	= **billabong** (Australian). A river meander which has been cut off from the main channel by the normal processes of channel migration in a mature river.
oxidizing conditions	Strictly, any redox potential of > 0. Water supplies should be maintained under oxidizing conditions, both for health and for organoleptic reasons.
oxygen	*See* **dissolved oxygen**, or **sag curve**.

oxygen deficit   The oxygen concentration required to oxidize reduced species in solution. A measure of contamination (mostly organic).

ozone   (O_3) An extremely oxidising metastable oxygen molecule with a half-life in air of about 12 minutes (although this can be greatly extended at low temperatures). It has an odour threshold of about 10 ppb, and is used to disinfect water by oxidising bacteria and other aquatic microorganisms. In this respect it is up to 5000 times as effective as free chlorine. The contact time for ozone disinfection of water is less than half an hour. Unlike most water disinfectants, ozone leaves no residual smell or taste. The widespread notion that ozone is beneficial to health is untrue; in fact, prolonged exposure to ozone is somewhat dangerous; even 1 ppm is sufficient to cause coughing and chest pain.

P

packer   A sealing device connecting the base of a borehole casing to the top of a screen of smaller diameter. Permanent packers are made of either lead or flexible neoprene rubber. Water samples can also be obtained from specific horizons within a screen by isolating the pump between upper and lower inflatable packers.

paddy   Throughout most of the world, and particularly in Asia, rice is cultivated in flooded fields, or paddies. Other crops are occasionally grown in paddies but it has become customary for 'paddy' to imply rice farming. During the growing period the optimum depth of flooding in paddies is typically ~50 to 70 mm, but farmers commonly err on the side of caution, flooding to >100 mm. From the perspective of water accounting it is important to include water used in paddy preparation. *See* **presaturation**.

palaeohydrology A relatively new multi-disciplinary interest which seeks to reconstruct prehistoric hydrological processes. So far, most palaeo-hydrological studies have been of mid-latitudes, extending back to about 50,000 BP, and have been more descriptive than quantitative. Palaeohydrology is distinct from dendroclimatology. Palaeohydrologic studies have influenced main-stream hydrology by estimating the upper limits for all flood envelopes.

palaeokarst Any karstic terrain which has been buried, and in which active dissolution has ceased. In some cases, erosion exhumes a palaeokarst environment, and active dissolution resumes.

Palmer drought index *See* **PDSI**.

pancake ice Most pancake ice occurs in the circumpolar oceans. In rivers pancake ice tends to accumulate in low-velocity pools as disks of ice up to about 25 cm diameter. In central Asia mid-winter pancake ice can be a metre or two in diameter, and loud grinding noises can be heard in the early morning ice release.

pan coefficient The multiplier which is required to convert the evaporation observed in an evaporation pan to that which would take place in a large water surface (lake or reservoir). The discrepancy arises because the humidity above a large lake suppresses high rates of evaporation. Values for a 'class-A pan' typically vary between about 0.6 and 0.7. For larger, non-standard evaporation pans, the larger the diameter, the closer the pan coefficient approximates to 1.0. In desert lakes where nocturnal advection is often minimal, the pan evaporation is almost entirely a day-time process.

P and M end-points The end-points of a carbonate / bicarbonate alkalinity titration. P is the phenolphthalein end-point (pink to colourless, at about pH 8.3), and M is the methyl orange end-point (yellow to pink, at about pH 4.3). Both end-points are approximations as the true end-point is also a function of temperature and ionic strength.

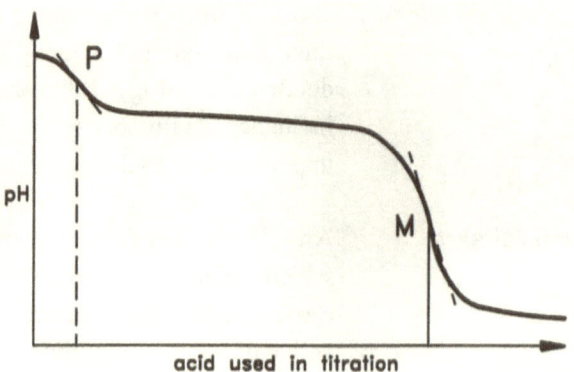

parametric data A statistical set of data that has a probability distribution assumed to be normally distributed. An example is the measurement error distribution of most physical parameters.

parasitic losses = **excess drawdown**; *q.v.* The turbulent head losses which occur in production wells as a result of high screen inlet and formation approach velocities.

partial area runoff Runoff from a storm that only affects part of a catchment.

partial penetration A borehole whose screen is only open through part (generally the upper part) of an aquifer.

partial series = **partial duration series**, = **peaks over threshold series** (POT). A series of maximum data points, usually taken from a hydrograph, in which all the lesser maxima - below an arbitrary threshold - are ignored.

Hence for flood studies, it is only the major flood peaks (the partial series) which are of interest, and which need be statistically treated. *cf.* **annual series**.

partition coefficient If the ratio of a solute or contaminant A in B is denoted as AB^R then the partition coefficient between any two phases such as liquid and solid, or two immiscible liquids, is $\dfrac{AB^{R_l}}{AB^{R_s}}$. Some authorities regard the term 'partition coefficient' as obsolete, preferring the terms 'partition constant' or 'partition ratio'.

passivation As the water's redox potential increases, there is increasing tendency for a metal (such as a steel screen, casing, pipe-work, pump, water tank etc.) to corrode. By increasing the potential still further, the metal oxidizes so effectively that a film of oxide (< 100 Å thick) prevents further oxidation. This process is passivation. In theory, passivation can also occur by the formation of a hydroxide film under strongly reducing conditions, but such negative redox seldom occurs in natural waters.

pathogens Disease-causing micro-organisms, many of which are aquatic. Most are bacterial, but they may also be viruses, yeasts, helminth worms, algae, cyanophytes, or protozoans.

PCA Principal Component Analysis. A powerful statistical tool for sorting out the relative importance of co-variables. It is a form of multivariate analysis, and is used, for example, in geomorphology to derive and calibrate empirical equations.

PCBs Polychlorinated Biphenyls. A group of environmentally stable, mutagenic and carcinogenic organochlorines, which have very low solubility in water. PCBs are acutely toxic to aquatic organisms, and because of their high solubility in lipids they bio-accumulate upwards throughout the food chain. Except for rigorously contained applications, PCBs were banned in most countries from about the 1970s, or by the Stockholm Convention (2001) on 'Persistent Organic Pollutants'. PCBs are amongst the most difficult **DNAPL**s (*q.v.*) to remove from aquifers, and continuing groundwater pollution arises from their past inappropriate disposal in landfills.

PDA = Personal Digital/Data Assistant, *alias* palmtop computer. PDAs are robust pocket-sized screen-touch devices, primarily for downloading field data to user-specified tables and complex formats. PDAs generally have a lot of 'bells and whistles', typically associated with high-end mobile phones, such as a GPS facility, camera, texting, e-mail, and Bluetooth connection to office computers, field instrumentation, *etc.*
In conservative countries do not confuse PDA with 'Public Displays of Affection'.

PDB The primary reference standard for carbon isotopes, and a secondary standard for oxygen isotopes. It is no longer available, but it had a ^{13}C abundance of 1.111%, and was obtained from the calcareous rostrum of a Cretaceous belemnite from the Pee Dee formation of South Carolina. This bizarre choice of standard was based upon PDB's similarity to the mean isotopic composition of marine limestone.

p.d.f. Probability density function. A smoothed graph of dimension-less magnitude *versus* frequency. They are widely used in flood and drought frequency analyses, and are the basis for a variety of different probability distributions. Errors from sample bias

and non-stationarity of data should be considered in the application of any p.d.f. The best p.d.f. to apply to a particular catchment is subjective.
A typical hydrologic p.d.f.

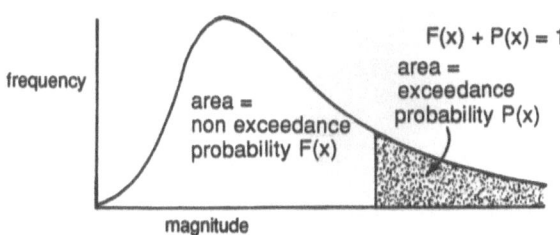

PDSI Palmer Drought Severity Index, *alias* 'Palmer Drought Index', is an American water-balance-based index for estimating long-term drought intensity, considered over months rather than weeks. The index is based upon precipitation and temperature, relative to normal. Numerically, the PDSI is normal at 0, mild drought is about -2, severe drought is about -3 and extreme drought can be as low as -5. Wetter than normal spells yield a positive PDSI. The PDSI has been criticized for yielding poor results in extreme climates and mountainous regions, and also for utilizing somewhat arbitrary assumptions. Nevertheless, it seems to work well, at least in America. Although the PDSI is conceptually simple, its calculation is rather tedious, so computer codes are available for the necessary number crunching.

peak storage A transient storage volume of a reservoir, between maximum flood stage and lowest intake.

Pearson type-3 A three-parameter probability density function (and associated cumulative density function) of the Γ family. It is much used in estimating flood probabilities from long discharge records, assuming that each flood record is independent and part of a

peat   homogeneous dataset. There are doubts about the whole rationale for using such p.d.f.s to estimate extreme events, but as probability models go, this is one of the best. *cf*: **gamma distribution**.

peat   A wholly organic accumulation, under reducing conditions, of buried vegetation (usually from a bog). Amongst several notable characteristics, peat has a very high porosity, low pH when wet, hydrophobicity (water repellence) when dry, a large ion exchange capacity, and a high calorific value. Peat is one of the most hydrophobic of all soils, and can lead to overland stormflow with only a few percent of stormwater wetting the top few cm of peat. Peat is estimated to cover 4 million km² of the land surface, and accounts for roughly 40% of total soil carbon storage. The hydrology of peat-land is complex, with some cases attenuating flood runoff, whilst other cases increase flooding. The management of peat-land is one of the most important linkages between hydrology and greenhouse gas emissions.

Peclet number $P_e = \dfrac{V}{D_a}$ where V is the velocity of groundwater flow, D_a is the coefficient of diffusion (as in the solute transport equation). P_e typically varies from about 10^{-2} to 10^8. When P_e is low, solute transport is dominated by diffusion, (it occurs in all directions), whereas larger P_es are characteristic of advective transport. Similarly, the Peclet number finds application in modelling heat flow in a porous medium, in which case it describes the relative importance of conductive (low P_e) and convective (high P_e) processes.

pedology   The study of soils.

pelites Clay-rich rocks: generally aquicludes, including shale, mudstone and schist. From Greek 'pelos' = clay.

pellicular water = Adhesive water. After gravitational drainage, the residual water, in the form of a thin film, which adheres to particles in a granular medium.

pendant ≈ **deckenkarren** (pillars of rock), ≈ **deckenzapfen** (ceiling cones of rock). An endokarst feature in which an anastomosing channel is divided by a residual pillar or roof cone.

pendular saturation Given a matrix saturated with two fluids (such as oil, air, water), if fluid 'A' is held at points of intergranular contact, then it is in a state of pendular saturation. Conversely, if 'A' is held as globules in the interstices (such as water globules within an oil-wetted matrix), then it is in a state of insular saturation.
NB: total porosity = insular + pendular saturation.

Penman equation One of the most widely used deterministic equations for the estimation of non-limited evapotranspiration from a vegetated land surface. There are two terms in the numerator, which describe the energy balance and advective components of the process. The notation varies greatly, but one of the simplest versions of the equation is: $E_p = \dfrac{B.H + E_a}{B+1}$ where E_p is the evaporation in mm, H is the available heat energy, E_a is an advection term: $f(v).(e_a - e_d)$ i.e. a function of wind run, and the saturated vapour pressure of water at the mean air temperature and dew point respectively. B is the Bowen ratio, $= \dfrac{\Delta}{\gamma}$ where Δ is the slope of the curve of saturated vapour pressure compared against temperature, and γ is the psychrometric constant.

Penman-Monteith equation A somewhat cumbersome but popular variant of the Penman equation which is now, arguably, the standard method of calculating total evapotranspiration. Penman's equation doesn't take vegetation into account, so Monteith modified the denominator of Penman's equation to take account of 'canopy conductance'. The equation is:

$$ET = \frac{\Delta(R_n - G) + \rho_a \cdot c_p \cdot (e_s - e_a)/r_a}{\left[\Delta + \gamma\left(1 + \frac{r_s}{r_a}\right)\right] \cdot \lambda \rho_w}$$

Where Δ is the slope of the saturated vapour pressure *versus* temperature curve, R_n is the net radiation flux at the surface, G is the sensible heat exchange from the surface to the soil (negative if the soil is cooling), ρ is the air/water density, C_p is the specific heat of dry air, e_s and e_a are the saturated and actual vapour pressures of air just above the surface, r_a is the aerodynamic resistance to turbulent heat transfer from the surface, r_s is the resistance of vapour flow from within the vegetation to the exposed vegetation canopy, λ is the latent heat of vaporization, and γ is the psychrometric constant.

Fortunately, there is software available to make this calculation much less formidable than it looks.

perched groundwater (or perched water table). A saturated groundwater 'lens' held by a localized impervious horizon within the unsaturated zone, and above the normal water table. Most alluvial aquifers contain small perched groundwater bodies, but they are seldom of significant volume.

perennial Permanent, perpetual, hence always flowing (*Latin: perennis*—through; *annus*—year).

period of oscillation ... of a seiche ... In a lake of effective fetch L_e, and mean depth D, the period of oscillation following wind set-up is given by: $T_n = \frac{1}{n} \cdot \frac{2Le}{\sqrt{gD}}$ where g is gravity, and n is the number of nodal axes in the oscillation.

peristaltic pump A low-discharge pump with very precise flow rates: often used for auto-sampling, analytical and metering devices.

permafrost Permanently frozen subsoil, usually saturated. The thickness of permafrost in polar regions may reach 400 meters.

permafrost table The maximum depth to which surface soil thaws above permafrost. Shallow perched groundwater is often controlled by the permafrost table configuration. The permafrost table is strongly influenced by exposure, surface water and aspect.

permeability Strictly, the intrinsic or specific permeability, k, (although nearly everyone just refers to 'permeability'). k is an intrinsic measure of the ease of fluid flow through a porous medium, and is independent of the saturating fluid properties. It has units of m², but as values are very small, they are often quoted in m² x 10^{-12}. In groundwater, common practice is to use the more convenient 'hydraulic conductivity', 'K', which is related by

$$= \frac{K\mu}{\rho g},$$ where μ is the fluid viscosity, and ρ is the fluid density, both temperature-dependent.

permeameter = **Oedometer cell** in soil physics jargon. A device for measuring hydraulic conductivity, usually of a cylindrical core's field sample. Permeameters can be of either constant head or falling head design. The principle (Darcy's law) is trivial, but there are several precautions required for good results, such as the avoidance of bubbles, minimizing 'edge effects', and 'undisturbed sampling'.

persistence The tendency for self-perpetuation of a process, such as rainfall or drought. More specifically, considering the auto-correlation of a time series, such as a rainfall moving average, or drought index; the persistence is the recession constant of the auto-correlation. Commonly, high persistence occurs during summer and winter, and low persistence during spring and autumn.

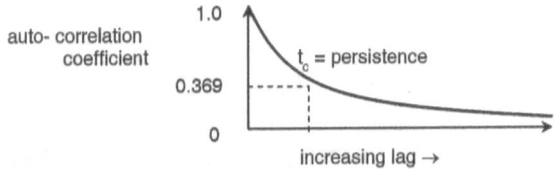

pervaporation One of several 'membrane technologies' used in the desalination of brine or brackish water (and also in some other chemical purification processes). A semi-permeable membrane separates brine on one side from a partial vacuum or 'sweep gas' on the other side. As water vapour diffuses through the membrane, it is collected and condensed as pure water. Note that pervaporation is a form of vaporization controlled by membrane permeability and salinity rather than by the thermodynamic equilibrium of a free evaporating surface.

pesticides Over a thousand pesticide compounds are in use around the world, most of which degrade rapidly. A problem exists with the longer-lived toxic and carcinogenic pesticides which find their way into the water cycle, and thereby facilitate bio-concentration, especially in the fatty tissue of higher animals, including humans. The maximum permissible concentration in drinking water varies according to individual compounds, but is typically about 0.01 mg.l^{-1}. The pesticides of most concern include aldrin, atrazine, chlordane, dacthal DDT, dieldrin, heptachlor, and lindane. (*See also* **POPs**).

pH The negative logarithm of the hydrogen-ion concentration in solution. In practice, the pH range varies from 0 to 14, with pH 7 taken as neutrality (although values outside this range are theoretically possible). Extreme ranges in nature vary from about 0.5, such as dilute sulphuric acid springs in Yellowstone Park, to 12.3 in hydroxide springs seeping from peridotite rocks. However, the vast majority of natural waters lie within the range 4.5 to 9.5. Rainfall normally dissolves sufficient CO_2 to depress the pH to about 5. Waters in contact with natural carbonate, such as limestone, equilibrate at about pH 8.5. The pH of waters strongly affect other aspects of the water's chemistry, and is arguably its most important chemical property. The pH is slightly temperature-dependent. It is normally measured electro-chemically, or approximated colourimetrically.

phenol A hydroxylated benzene ring molecule, with numerous related compounds. Natural phenol concentrations in ground-water are less than 1 µg.l^{-1}, but higher concentrations are common from industrial and agricultural wastewaters. The odour threshold of phenol in water is about 0.3 mg.l^{-1}, whereas the taste threshold is about 8 mg.l^{-1}. *Chlorinated* phenol compounds have detection thresholds in the order of thousands of times less than those of pure phenol.

phenolphthalein   *See* **P and M end points**.

phosphate   Phosphorus occurs as a variety of phosphate ions (such as H_3PO_4, $H_2PO_4^-$, HPO_4^{2-} and PO_4^{3-} according to pH) and other compounds, both soluble and particulate, and is a key nutrient in both agriculture and water quality. Natural phosphate concentrations are typically about 10 to 30 $\mu g.l^{-1}$. Much higher concentrations can arise as runoff from fertilized farmland and stockyard wastes, or from detergents, optical brighteners, sewage, industrial effluents, and occasionally from hydrothermal systems.

photic zone   The upper layer of a water body which absorbs 99% of the incident surface insolation. The thickness of the photic zone varies from a few cm in the case of high sediment loads or eutrophic blooms, to about 40 metres in very clear water. It is also slightly wavelength-dependent.

photosynthesis   The basic plant growth mechanisms in which energy (light) + CO_2 + H_2O → carbohydrates. The water so consumed is less than about 1% of the total water transpired by the plant.

phreatic decline   Groundwater recession; the natural fall in dry season piezo-metric head.

phreatic divide   A 'groundwatershed' or subsurface catchment divide, which may or may not coincide with a surface water divide. It is often impossible to locate the phreatic divide accurately in areas of low- or ill-defined hydraulic gradients, or beneath sand dunes or tundra.

phreatic zone   Also known as **Zone of saturation**. The saturated part of a water table aquifer.

phreatophyte	A class of vegetation, common in arid areas, which is capable of tapping the water table, often to substantial depths (to tens of metres). *cf.* **xerophyte**.
phytokarst	= Biokarst. Exokarstic features caused by natural organic acids or by inorganic acids of biological origin. Examples include solution pans beneath vegetation and soil where bacteria increase CO_2 in the root zone. Coastal and lakeshore phytokarst results in extremely jagged, pitted and carious limestone surfaces.
phytometer	A vessel for measuring 'whole plant transpiration', in which moist soil is sealed, except for protruding vegetation.
phytoplankton	Phytoplankton are autotrophic microorganisms, which are present in all aquatic environments, both freshwater and sea-water, including some of the most hostile environments, such as thermal springs and Antarctic lakes. They may be unicellular or filamentous. In drinking water they are mostly harmless but some exude a bad odour, and a few are seriously toxic. At least 30,000 species are known, and probably far more have yet to identified. Of these, several dozen 'usual suspects' can pose problems in reservoirs, requiring treatment with algicide. Annual cycles of phytoplankton ecology are caused by seasonal changes in temperature, sunlight, stratification and nutrient availability. For example, in one reservoir with a Mediterranean climate, winters (<15°C) may be dominated by the diatom *Cyclotella*, whereas in summer (>15°C, stratified, higher nutrients), may be dominated by such green algae as *Oocystis*, *Dictyosphaerium* and *Ankistrodesmus*.
Piche gauge	A simple evaporimeter consisting of an inverted graduated cylinder with a porous base from which evaporation takes place. It is not as accurate, or as

spatially representative, as a class A pan evaporimeter, but correlates reasonably well with the second (advective) term in the numerator of the Penman and Penman-Monteith equations for evaporation.

piedmont Literally 'the foot of a hill or mountain'. Relatively flat alluvial fans and terraces found at the edge of mountainous areas, and which are characterized by a marked change in hydraulic gradient and fluvial style. The piedmont may exhibit a change from erosion to deposition, with medium to very coarse-grained sediments. Hydraulic conductivities and SWL fluctuations tend to be large in this environment.

piedmont glacier A glacier that extends beyond its valley confines to spread out into a piedmont zone.

piestic interval The differential pressure between two isopotentials.

piezometer A narrow tube, pipe or borehole for measuring the moisture in a soil, water level in an aquifer, or pressure head in a tank, pipeline *etc*. It may vary from a glass tube and manometer in a bench top experiment, to a large-scale borehole in a deep aquifer. The end in a saturated medium may vary from a simple open section to an elaborate screen and gravel pack. Where the water velocity is significant, or where waves occur, the piezometer may have to be installed in a stilling basin. Inside diameters vary from about 7 mm to 8 cm.

piezometric surface Alias the **atmospheric pressure surface**, or the potentio-metric surface, = the water table in an unconfined aquifer. The notional 'static water level' that would occur under unconstrained conditions, such as observed in boreholes.

piezometry The process whereby a series of water level measurements from an array of narrow bore observation bores (minimum three, occasionally hundreds) is used to determine the groundwater pressure contours, and thereby determine the direction and hydraulic gradient of flow. Piezometry is the requisite basis for conceptualizing, initializing and calibrating groundwater models. You may come across idiot clients who think you can go straight to modelling without the need for piezometry.

pingo = Hydrolaccolith, = cryolaccolith, = frost heave. A small hill of up to about 40 metres' elevation (usually much smaller) caused by an inflow of groundwater between a frozen ground surface and permafrost, followed by continued freezing.

pipe / pipeflow Pipes are natural conduits which develop in steep hillsides in soils overlying low permeability bedrock. Pipeflow is a form of interflow which drains hillsides very efficiently, thereby resulting in a very rapid runoff response in river headwaters.

Piper diagram A developed double trilinear (isometric) diagram to portray the relative proportions of the major ions. The absolute ionic concentrations can be represented by symbol size.

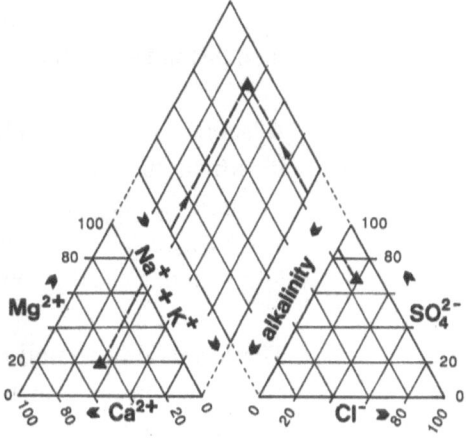

piping If groundwater seepage beneath a dam, or within an earth dam, reaches sufficient velocity to dislodge particles, then progressively enlarged seepage paths, or 'piping' can lead to catastrophic failure. In fact piping is the single most common reason for dam failure, accounting for about a third of all dam breaches. Hence an aim of dam design should be to drain excess pore pressure by creating a sufficiently low hydraulic gradient that piping is not initiated. It is usual practice to estimate the exit hydraulic gradient (by solving the Laplace equation), in the light of the prevailing soil properties.

piracy A natural phenomenon by which water from one surface catchment transfers to another catchment by groundwater flow. This is most common in karstic terrains.

piston flow A process in which new recharge, entering the start of a groundwater pathway, is in hydraulic continuity with the rest of the pathway, thus permitting the transmission of hydraulic pressure throughout the entire pathway. Consequently, the new recharge displaces old water at the discharge end of the pathway. There has been much controversy over the validity of this concept in many environments. The hydraulics of most catchments are more complex than piston flow alone.

Pitot tube A laboratory instrument for measuring flow velocity, in which the difference between the static and dynamic heads, Δh, is related to the velocity by: $v = c\sqrt{2g.\Delta h}$. The constant C corrects for head losses, and is usually about 0.98.

pitting The formation of small cavities on metal surfaces, either by corrosion or by cavitation. The latter is particularly damaging to rotodynamic devices.

pixel	The smallest resolvable unit of a digitized picture, of uniform spectral signature. For example, a Landsat image may have a pixel footprint of 120 metres x 100 metres, and a spectral signature defined by intensities at 7 wavelengths. A given satellite scene may be composed of millions of computer-processed pixels.
planimeter	A remarkably simple instrument for measuring areas from a map. With practice it can be used to an accuracy of 1 or 2 %. It is particularly useful for estimating catchment areas with very complex boundaries.
plasma chemical reactor	An experimental but potentially effective method of killing bacteria and destroying some POPs (such as 2,4-D). The reactor uses multi-electrode sparks in a water column to generate UV radiation, ozone and hydrogen peroxide transients.
plastic limit	P_w: The water content of a soil or clay when it just begins to crumble when rolled into a thread of about 3 mm diameter. It is one of the **Atterburg limits** *(q.v.)* and is a measure of the water content that can be held with some degree of soil rigidity. Some skill is required for its consistent and accurate measurement. Normal clay values vary from about 26% for kaolin to about 76%, but Na-smectite may have a P_w value in excess of 100%. The adsorbed cation is at least as important a variable as the clay species.
plasticity index	$I_p = L_w - P_w$. It is the range of soil / clay moisture over which wet soil bearing strength is partly lost, but in which some residual rigidity remains. *See* **Atterburg limits.**
plate count	Also known as 'total plate count' or 'standard plate count'. A measure of bacterial pollution of water. It

uses the incubation of a diluted water sample on a nutrient agar in a petri dish, usually at 22°C or 37°C, for 24 hours. Dilutions range from 10 to 10,000-fold. The cultured colonies are a mm or more in diameter, and can be counted by eye or, in the case of the more modern fluorescent methods, by automation.

playa   = Salar, = dry salt lake.

plotting position   When analysing a time-series of data, such as for flood-estimation/projection the convention is to assign some measure of exceedence probability, the 'plotting position', to a partial or annual series. The plotting position is purely a function of the number of data points in the series, 'N', and the ranked order in the series, 'm' and is independent of the actual data magnitudes. That is, for the largest value in the time series, m=1, whilst for the smallest, m=N. All plotting positions are approximations for which there is no perfect or ideal definition. Whatever you choose makes little difference, and is much less than the errors of data measurement. Take your pick from the following suggested expressions for 'plotting position':

Landwehr	$\dfrac{m-0.35}{N}$	Weibull	$\dfrac{m}{N+1}$
Tukey	$\dfrac{3m-1}{3N+1}$	Chegodayev	$\dfrac{m-0.3}{N+0.4}$
Cunane	$\dfrac{m-0.4}{N+0.2}$	Blom	$\dfrac{8m-3}{8N+2}$

plume   The configuration of a diffused and advected pollutant in either the atmosphere, surface water or groundwater.

pluvial Pertaining to a particularly rainy period. In palaeohydrology, a pluvial specifically means a long wet period in a relatively mild climate contemporaneous with glacial epochs in colder regions.

pluviothermic quotient Q, One of the lesser-used indices of aridity, defined as $Q = \dfrac{100P}{M^2 - m^2}$ where P is the mean annual precipitation (mm), M is the mean maximum temperature of the hottest month (°C), and m is the mean minimum temperature of the coldest month (°C).

pluviometer A device for recording both rainfall increments, and their time of occurrence, hence yielding rainfall intensities. Formerly auto-graphic, but now designed with solid-state data loggers. The graphical output is the pluviograph.

point precipitation Some of the most serious errors in hydrology come from extrapolating from 'point precipitation', such as from a raingauge, to gauge-catchment size, typically hundreds of km². That is, an areal extrapolation in the order of 10^{10}. The problem of point to areal extrapolation becomes even more acute when applying climatic data to global climatic models.

Poisson's equation This may be regarded as a variation of the linearized steady-state equation, and takes account of vertical recharge to an aquifer in an otherwise horizontal groundwater flow régime:

$$\frac{\partial^2 h}{\partial x^2} + \frac{\partial^2 h}{\partial y^2} = -\frac{R_{x,y}}{T}$$ where $R_{x,y}$ is the mean recharge rate over the $\Delta x, \Delta y$ element being considered. The Poisson, Boussinesq, Diffusion, and Laplace equations all come out of the same, stable' (conservation of mass), together forming the basis of

analytical and numerical modelling of ground-water flow. The complete groundwater flow equation is:

$$T_x\frac{\partial^2 h}{\partial x^2}+T_y\frac{\partial^2 h}{\partial y^2}+T_z\frac{\partial^2 h}{\partial z^2}+R_{x,y}+\frac{K'}{B}(h_a-h)=S\frac{\partial h}{\partial t};$$

where h is the hydraulic head, K' is the vertical hydraulic conductivity of a confining layer, B is its thickness, R is the recharge rate, S is the aquifer storativity and T_n is the hydraulic conductivity in the n direction. In a homogeneous aquifer the equation may be recast in radial ('polar') form.

polarization A natural ground response to direct current which interferes with resistivity measurements. It is largely overcome by using low- frequency alternating current, and 'porous ceramic' electrodes.

polar structure The highly asymmetric electrical charge distribution of water molecules, in the form of a 'quadrapole', is largely responsible for water's extraordinary chemical and physical properties. (Compared to other chemicals, water is anomalous in almost every respect). Polarization, or lack of it, is also the main factor in determining the solubility of other liquids in water.

polder An area of reclaimed land below an adjacent water body. The Dutch polders are the classic example.

polje A large flat depression in limestone caused by karstic erosion to a fixed base level. Poljes are tectonically related insofar as dissolution initiates in and around fracture sets.

polymictic A polymictic lake overturns and mixes frequently due to large thermal fluctuations, and consequent

Dictionary of Hydrology and Water Resources

instability of its thermo-cline. This is most characteristic of high altitude tropical lakes.

polysaprobic conditions Highly polluted rivers or lakes, exhibiting euxinic reducing conditions.

ponded water = depression storage.

pool level The water level in a reach of a river under regulated flow conditions, or the level in a reservoir. In the latter case 'normal pool level' is that of the spillway crest. In a long reservoir with throughflow, the water surface, and hence local pool level, is not horizontal.

pools Deep depressions in river beds.

POPs 'Persistent Organic Pollutants' generally refers to harmful chemicals, of low aqueous solubility, that degrade slowly or not at all, in the environment. Many pesticides and halogenated organics figure prominently in this long list of chemicals. Also included are PVC, various solvents, heavy metals and their organic salts, and endocrine disruptors ('gender benders'). Of greatest concern are the so-called "dirty dozen" comprising: aldrin, chlordane, DDT, dieldrin, dioxins, endrin, furans, heptachlor, hexachlorobenzene, mirex, PCBs and toxaphene.

The Stockholm Convention, implemented in 2004, and managed by UNEP, is a near-global treaty which requires all signatory countries to eliminate or restrict the release of POPs into the environment. The convention is very wide ranging, including clean-up of contaminated water supplies, phasing out of production, dismantling of old ships, *et cetera*.

porosity The proportion of space between the particles of a sediment, variously denoted as: a, n, α, ε, or η. It may be fractional, but is usually quoted in percent: $= V_v/V_t \cdot 100$, where V_v is the volume of voids, and V_t is the total volume of the porous medium. Engineers sometimes express porosity in terms of, e, the 'voids ratio' $e = V_v/V_s$, where V_s is the volume of the solid fraction. Since $V_s + V_v = V_t$, $a = \dfrac{e}{1+e}$.

post-audit Evaluation of the accuracy of predictions by previously 'validated' computer models. (It is often a salutary experience for those who have a naïve faith in computers!)

potable water Water that is acceptable to drink according to chemical, biological and aesthetic criteria. For example, water that conforms to published WHO, EU, or USEPA standards, or to national standards derived therefrom. Potability is defined scientifically, not by popular perception.

potamology A branch of hydrology concerned with streamflow.

potassium (K). Normally the least of the major cations in solution. Potassium is geochemically about as abundant as sodium, but K feldspar breaks down very slowly, whilst K^+ in solution is readily adsorbed onto clays. Its typical range in surface and groundwaters is about 0.5 to 20 mg.l^{-1}. The chemistry of K and Na is similar, with K salts being slightly more soluble, so K salts are amongst the last to be precipitated from an evaporating brine.

potassium chloride The electrolyte used in some electrodes, and the preferred salt for calibrating E.C. meters. Some standard values are:

molar KCl concentration	E.C. in mS.cm⁻¹ @ 25°C
0.001	147
0.005	718
0.01	1413
0.02	2767
0.05	6668
0.1	12900

potential (1) *See* **hydraulic potential**.
(2) Electrical potential is relative, and is measurable in ± volts. The reference, such as for Eh measurement of natural waters, is a 'platinum half cell'.

potential evaporation The amount of evaporation that would occur from a wet surface, given an unlimited water supply. It is a function of the micro-climatic energy balance.

potential evapotranspiration The maximum evaporation and transpiration that could occur in a non-water limited environment. *cf.* **p. evaporation**.

potentiometric surface = Piezometric surface, = the pressure surface. In an unconfined aquifer the potentiometric surface *is* the water surface: in a confined aquifer, the potentiometric surface is an hypothetical surface *above* the actual water surface.

precipitable water = precipitable moisture $W(mm) = \dfrac{0.1}{g} \int_{P_2}^{P_1} q.dP$

where q is the specific humidity (g.kg⁻¹), and P is the pressure in HPa. Values vary from about 60 mm at the equator, to almost zero at the poles. The precipitable moisture is the total moisture in a static column of the atmosphere, of unit cross-sectional area. All of this moisture is never actually precipitated, although tropical storms and extremes of uplift might precipitate over 90% of W. Naturally, advection allows precipitation far in excess of W at any given point. Precipitable moisture above about 250 HPa is negligible.

precipitation General term for rain, snow, hail, frost and dew. Also the water equivalent expressed in mm. Precipitation can be regarded as the input to a catchment, which is subsequently broken down into the components: evapotranspiration + quickflow + baseflow + infiltration + inter-basin transfer, *etc.*

precision The degree of exactness to which repeated measurements can be made. It has been variously defined as the coefficient of variation, or as three times the standard deviation of the background signal during measurement. It is an absolute quantity, as opposed to 'accuracy', with which it is sometimes confused.

preconsolidation Plastic sediments, mainly clays, that have previously been subjected to higher than existing pressures, are pre-consolidated. Significant amounts of water will only drain from this sediment once the previous highest pressure has been exceeded. *cf.* **critical head**.

presaturation In traditional paddy cultivation it is common to soak the land by flooding for three or four weeks, prior to transplanting rice seedlings from the nursery. This pre-flooding practice is to (i) control weeds, (ii) maximize field capacity - minimizing wilting loss, (iii) leach out any salt build-up in the soil, (iv) increase soil nutrient mobility, and (v) to maximize the rice yield.
Presaturation can be wasteful of the water resource in water stressed areas, but the duration of presaturation can be reduced by mechanized tilling.

pressure aquifer = confined aquifer = artesian aquifer.

pressure chamber Equipment for measuring xylem and leaf water potentials (suction). The pressure within the chamber is raised, by injecting gas, until a cut plant or leaf begins to exude fluid, at which point $\Psi = -P_{gas}$.

pressure head ρ.g.h. or specific gravity, gravitational constant, head.

pressure membrane Apparatus for measuring the soil moisture potential (suction) of a soil, in much the same way as a pressure chamber.

price elasticity A dimensionless measure of consumer sensitivity to the change in price of a commodity i.e. The change in price ÷ the change in quantity, or: $\varepsilon = \frac{\Delta Q}{Q} \cdot \frac{P}{\Delta P}$. ε is considered 'elastic' if $|\varepsilon| > 1$, and inelastic if $|\varepsilon| < 1$.

primary porosity Inter-particle void space originating from the time of deposition.

primary salinization Natural salinization.

priority pollutants The 130 most important (toxic and/or common) pollutants associated with industrial effluents, as defined by the USEPA.

probability of failure (of a reservoir supply), $P_f = \frac{\Sigma S}{\Sigma N}$, where ΣS is the cumulative time for which there is no active storage, and ΣN is the un-restricted cumulative time for which water release is available. Conversely, the *reliability* of supply = 1 - P_f.

probable maximum precipitation Other factors being equal, the PMP gives rise to the probable maximum flood (PMF). It is an untestable hypothetical value resulting from a 100% efficient storm, under saturated antecedent conditions.

production well = **production bore** (hole). The pumped well in an aquifer test, as opposed to observation wells. A wide-bore well, fully developed and screened for water supply, drilled on the basis of previous exploration bores.

prokaryotes One of the five-fold classification of life forms comprising bacterial and cyanobacterial microorganisms.

propellor *See* **impellor**.

protozoa A class of micro-organisms ubiquitous in soils, and common in natural waters. Most are harmless, but a few enteric protozoans, ingested from fecally contaminated water, cause mild to severe gasteroenteritis. Examples include *Cryptosporidium parvum*, *Entamoeba*, and *Giardia* (causing cryptosporidiosis, amoebic dysentery, and giardiasis respectively). Infection is usually caused by drinking water contaminated by oocysts (protozoan eggs), which burst upon contact with stomach acids, spreading parasitic organisms through the intestines. However, the rare *Naegleria fowleri* infects through the nasal membrane during swimming in contaminated water. Naegleria causes severe amoebic meningitis which is nearly always fatal. Some less virulent amoebae are also health hazards, but, protozoans also play an essential ecological role by consuming algae, fungi and bacteria, both in waste-water treatment and in soils.

pseudokarst Collapse features resembling those of karst, but caused by processes other than dissolution.

psychrometer An instrument for measuring air or soil humidity. *See* **sling psychrometer.**

psychrometric constant Occasionally referred to as the 'hygrometric constant'. This is an expression for estimating the actual vapour pressure from wet and dry bulb temperatures, as measured in 'met stations': $e_a = e_s - \gamma.(T_d - T_w)$, where e is the actual and saturated vapour pressure, and $T_{w,d}$ are the wet /dry bulb temperatures. It is also a factor in the Penman and Penman-Monteith equations for calculating evapo-transpiration.
γ, in kPa.°C^{-1}, is given as: $\gamma = \dfrac{(c_p).P}{\lambda_v.MWR}$ where c_p is the specific heat of air at constant pressure (MJ.kg^{-1}.°C^{-1}), P is the atmospheric pressure in kPa, λ_v is the latent heat of vaporization of water (MJ.kg^{-1}), and MWR is the molecular weight ratio of water vapour to dry air. At sea level the psychrometric constant is approximately 0.066 kPa.°C^{-1}, but the 'constant' is a bit of a misnomer. It may be reasonably consistent at any given location, but it varies a little with air composition, and a lot with altitude (pressure).

pulling casing Removing casing from an abandoned well, commonly by means of hydraulic rams.

pump Most large-scale groundwater extraction uses multi-stage electric submersible pumps. Depending upon circumstances other options include deep-bore turbines, centrifugal, jet (venturi) and draw-lift pumps. Manufacturers publish catalogues with all the depth-diameter-power-drawdown characteristics.

pumped storage A specialized form of Hydro-Electric Power (HEP) in which excess coal-fired or nuclear generating capacity during off-peak hours is used to pump water to a high reservoir. During peak demand, the pumps are reversed, and the same water is used to generate electricity.

pump power P_p = water power ÷ pump efficiency, η_p, ie:

$$P_p = \frac{g.Q.h}{\eta_p}$$ in units of N.m.s^{-1} where g is the specific weight of water, Q is the pumping rate, and h is the head gained. η_p is typically about 0.8. *cf.* **turbine power**.

pump-set At its most basic, a pump-set is an assemblage comprising engine/motor, drive-shaft and pump. Arguably, the pump-set may also include the fuel tank, electrical control board and rising main. It can be fixed, mobile or even floating. Most pump-sets are diesel, but they can also be petrol or electric driven. The pump may be suction or submersible. Mobile pump-sets are preferred for illegal extraction from irrigation channels, whilst fixed pump-sets are preferred for ground-water extraction. Pump-set efficiencies tend to be rather low, at <50%. There are 20 million pump-sets operating in India alone. For many farmers in developing countries their pump-set is their most expensive asset. Indian farmers often sleep next to their pump-set to prevent it from being stolen.

pump test = **pumping test**. Often a misnomer for an 'aquifer test'. (The usual objective is to test the aquifer, not the pump!) A means of stressing the aquifer by pumping from it, and matching the aquifer response to an analytical model. There are several different

Dictionary of Hydrology and Water Resources

techniques which may be either steady-state or transient.

PVC (Poly-Vinyl Chloride). One of several thermoplastics used to make well casing. It is lightweight, cheap, (relative to stainless steel) and chemically inert. It is mostly used for shallow wells where mild steel casing would corrode. However, it is not nearly as strong as steel. PVC or HDPE, ± clay is also used to seal landfill, or as an impervious wall against polluted groundwater flow. In the groundwater environment, PVC has an expected life of between about 75 and 100 years.

pycnocline [*Greek puknos; dense*].The stable layer of a stratified lake, characterized by a steep density gradient with depth. This may be due to temperature or salinity or both. It is possible, though unusual, to have several pycnoclines in the same lake, at differing depths, in which case there will be alternations of stratal stability and instability.

pycnometer A small glass vessel designed to hold a very precise volume, and which is used to measure the density of a liquid.

pyranine A hydrophilic fluorescent dye, $C_{16}H_7Na_3O_{10}S_3$, which changes from blue to green between pH 6.5 and 7.5. Pyranine fluorescence is occasionally used in soil-moisture tracing.

qanat = **kanat** (anglicised Persian), = **foggara** (north Africa), = **falaj** (Arabia), = **gallerias** (Mexico). A line sink dug upstream at less than the local groundwater's hydraulic gradient, thus intercepting the water table at depth at some distance from the outlet portal. A common misconception is that the water mainly originates from the most upstream shaft, or 'mother well'. In reality, it is the active section of the qanat which is most efficient at collecting groundwater. The technology is well over a thousand years old, and is still widely used for irrigation supplies in arid parts of Africa, Arabia and south-west Asia.

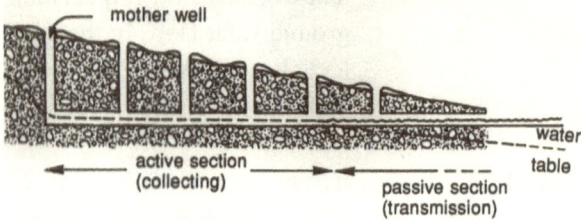

QA/QC — Quality assurance / quality control. Water quality laboratories require a management system to ensure consistency and accuracy of results, and to attain certification. This comprises chain-of-custody and other documentation, quality control samples (inter-laboratory comparison of results), standard operating procedures, and external quality assessment.

quality of snow — The amount (%) of liquid water in a snow sample, as determined calorimetrically. The free water content of snow normally lies in the range 0 to 5%, corresponding to a water quality of 100 to 95%.

quasi-biennial oscillation In the Stratosphere, above about 100 mb (hPa), and most notably higher than 10 mb (hPa), there is a zonal wind that oscillates between westerlies and easterlies with a frequency of about 20 to 36 months (mean of 28 months). These winds gradually descend from 10 to 100 mb. These winds derive their momentum from equatorially trapped Kelvin waves (westerly) and Rossby-gravity waves (easterly).

Normally there is considered to be a disconnect between stratospheric and tropospheric processes, but the phase and intensity of QBOs is now thought to be linked to rainfall in the Sahel, NW Pacific cyclone frequency, Atlantic hurricanes and monsoons.

quickclay Extremely sensitive uncompacted clays (sensitivity > 16) which virtually fluidize after minor mechanical disturbance or beyond a certain threshold of moisture content, sometimes with disastrous results. For example, some volcanic mudflows are known to have been triggered by rainfall, others by seismicity. Quickclay development is complex, but at least one explanation involves the leaching of Na^+ ions from smectites, resulting in weakening of interparticle water bonds, resulting in the loss of orientation in polarized water molecules.

quickflow The flood runoff, or fast response to rainfall of a stream or river. It comprises channel precipitation, surface runoff, and rapid interflow.

quicksand Sand in which the intergranular pressure $\leq$ the hydraulic pressure of upward flowing groundwater. The sand-grains are buoyant, and behave as a fluid.

quiet reach A straight reach of river, of more or less constant cross section, displaying uniform steady flow with minimum helicoidal flow. An undisturbed ideal reach in which flowlines are straight and parallel, and stream gauging is therefore relatively easy.

quoin = **quoin post**, = **heelpost**, = **hinge post**. The sturdy vertical post which supports each side of a lock gate.

R

race	A reach of rapid streamflow. It is a headrace or tailrace up-stream and downstream respectively from a mill, turbine, *etc*.

rack	A grille, usually just steel bars, placed across a stream, culvert or intake to 'sieve out' the coarser floating or suspended débris.

radar equation	(for estimating precipitation) $P_r = k \dfrac{\sum n.D^6}{\lambda^4 R^2}$

where P_r is the reflected power, k is a characteristic system constant, λ is the radar wavelength, R is the distance between transmitter and reflector, and n is the number of raindrops of diameter D. The drop diameter typically varies from about 0.5 to 5.0 mm, with larger drops being aerodynamically unstable. Obviously, with such sensitivity to D, the kind of rain is critically important to the radar interpretation.

radar wavebands	Weather radar is available in three wavelengths: X band (3 cm), C band (5.0 or 5.7 cm), and S band (10 cm). The shorter wavelength is more sensitive, and is capable of detecting very light rain or snow, but is heavily attenuated, and cannot 'see' to the far side of a rainstorm. On the other hand, the S band is well suited to heavy rain, and is therefore widely used in aviation.

radial well	A wide-bore well from the base of which several horizontal collector galleries radiate. It is complex and difficult to construct, but may be very appropriate in a heterogeneous, fractured, low 'K' aquifer.

radioactive decay law $C_t = C_0.e^{-\lambda t}$ where C_t is the radioactivity (or radionuclide concentration) at time t, C_0 is the initial radioactivity, and λ is a characteristic decay constant of the radionuclide in question.

radiogenic Resulting from radiation or radioactive decay. For example, Radon is a radiogenic trace element, sometimes used to detect deep fractures in an aquifer.

radionuclide = **radioisotope**. Any radioactive species.

radiosonde A balloon-lofted radio transmitter which conveys an atmospheric profile of pressure, temperature and humidity, from which the precipitable moisture (amongst other things) may be calculated.

radius of influence The minimum radial distance from a pumped well at which the drawdown is imperceptible; that is, the feather edge of the cone of depression. In theory, for a confined aquifer, $r_i = \sqrt{\dfrac{2.25 T t}{S}}$ in terms of transmissivity, storativity and time. This may reach several kilometres.

radius of investigation = radius of penetration. In borehole logging this is the effective radius to which a probe can sense a given parameter, such as $\gamma - \gamma$, neutron or resistivity. Ideally, a probe should measure the parameter of just the virgin zone. In practice, the parameter is measured within a volume described by the radius of investigation, which includes parts of the bore itself (usually filled with drilling fluid), the filter cake, the invaded zone and the virgin zone.

RAFT Rising Air Flotation Technique. A means of measuring the velocity profile of a stream by noting the horizontal surface displacement, L, of bubbles released from the stream bed. If the depth is d, and the terminal velocity of the bubbles is V_b, then the mean velocity at a given point is: $\overline{v} = \dfrac{L.V_b}{d}$

rainday Any 24-hour period between successive raingauge readings, in which a 'daily' raingauge registers ≥ 0.1 mm. (The minimum increment varies is a few countries).

rainfall excess
1) In isotopic studies, the δ deuterium value corresponding to $\delta^{18}O$ with respect to SMOW = 0. δD is the intercept of a precipitation line with the δD axis. Globally, the excess is about 10, but it reaches about 22 in the eastern Mediterranean region. 'Rainfall excess' is not a valid parameter in hot arid conditions due to non-equilibrium conditions.
2) In runoff estimation, the rainfall excess is that part of the total rainfall which contributes to quickflow.

rainfall excess hyetograph A histogram of effective rainfall with time for a given storm—generally an hypothetical design storm.

rainforest Conventionally defined as forest with a mean annual rainfall of >2500 mm. Nearly a third of the Earth's oxygen turnover is through the photosynthesis of rainforests. Tropical rainforests are concentrated in three regions: Central and South America, The Congo Basin, and South-East Asia from Myanmar to the Western Pacific Islands. Temperate rainforests are much smaller in extent. By 2012, 90% of the rainforest had been lost in West Africa and the Philippines. Near total loss of rainforest is also expected in Indonesia, Malaysia, Papua New Guinea, most of Madagascar, and up to 60% of the Amazon basin. The combined effects of climate change and rainforest clearance is expected to have profound effects upon the global hydrologic cycle.

'rain follows the plough' A notion from pioneering folk lore, from America, Australia and elsewhere, that land clearance for cultivation has the effect of creating a wetter climate. This contrasts with the scientific data that, if there is any relationship at all, it is probably the reverse effect.

rainfall intensity The rate of precipitation in $mm.h^{-1}$. In temperate climates the mean rainfall is about 1.25 $mm.h^{-1}$, whereas in the tropics it is about 15 $mm.h^{-1}$. However, the rate is extremely variable. Globally, the data on rainfall intensity is sparse compared to total rainfall. Intensity data from a given pluviometer is usually presented as intensity - frequency - duration (IFD) curves.

rainfall loss That part of the total rainfall which does not contribute to quickflow, but which is lost as interception, depression storage, evapotranspiration, and in making good the soil moisture deficit.

rainfall-runoff All rainfall-runoff relationships are gross simplifications of a very complex set of processes. The main controlling variables are rainfall amount and intensity, soil type and antecedent conditions, slope, geology, vegetation cover and uniformity, interception, depression storage, vegetation-growth-stage, and seasonality. There are numerous approaches to quantifying this important relationship, all of which should be viewed with a critical eye, and qualified by error analysis. Changing land-use and evolving climate require the rainfall-runoff to be periodically re-assessed.

raingauge There are dozens of raingauge designs, most of which are manually read funnel-shaped collectors draining to a storage sump. Aerodynamic flow around the rim of the gauge results in a systematic undercatch. Additional losses are caused by splash and evaporation. The overall error is typically -10 to -20%, but in exposed extremely windy conditions, the under-catch may be > 50%. Most published raingauge data implies an unrealistic precision and accuracy. *See also* **pluviometer**, **radar**, **weighing lysimeter**, and **optical rain-gauge**.

rain shadow Compared to adjacent areas there is a marked rainfall deficit on the lee side of a high relief area, caused by descending, and hence warming air. (Rainfall occurs by cooling the air).

rainwash The downhill migration of soil particles in response to the impact of raindrops.

rapid drawdown Rapid reduction in water level in a lake or reservoir. The resulting differential pressure causes bank instability, possible collapse, and sudden wave generation. Rapid drawdown has been the cause of some major hydrological disasters.

rating curve The graph of stage *versus* discharge. It is linear and unique under 'section control' conditions, but looped and variable for channel control conditions.

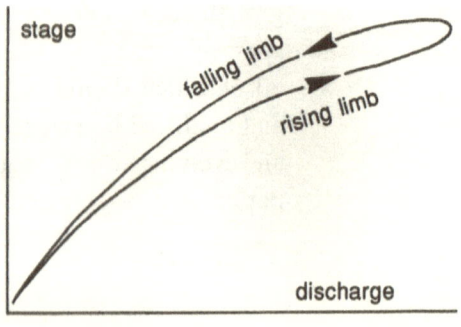

rating tank *See* **towing tank**.

rational method Estimation of the peak discharge rate of a catchment by means of empirical equations of the form: Q = aCIA, where Q is the peak flow rate, a is a constant for consistency of units, C is the dimensionless runoff coefficient, I is the average rainfall intensity over the time of concentration, and A is the catchment area in square kilometres. C and I (and hence Q) must be specified for some value of average recurrence interval.

raw water Untreated water.

Rayleigh distillation A model of the isotopic distribution of an air mass as it continues to precipitate rain. Since the heavy isotopes tend to condense more readily, the Rayleigh model predicts the relationship: $\dfrac{R}{R_0} = f^{(\alpha-1)}$ where R is the heavy /light isotopic ratio for residual atmospheric vapour, R_0 is the original isotopic ratio, f is the fraction of vapour remaining, and α is the isotopic fractionation factor.

reach A relatively short section of a river; such as between confluences, locks, or gauging points, or a continuous straight stretch, or a length of uniform flow, or of known storage characteristics (as used to calculate flood routing).

realignment Engineered change in a river's plan view, usually to accelerate flood discharge, such as by short-cutting a meander.

reamer A heavy drilling implement for straightening and enlarging a borehole.

recession The process of decline (such as of groundwater level or surface flow) after rain or artificial replenishment. Often approximates to an exponential decay.

recession constant The time taken for a stream discharge to recess by a factor of 1/e, which is to 0.3679 of its original value.

recession curve The 'tail' of a flood hydrograph—which often plots as linear on a log Q *versus* time hydrograph.

recharge Infiltration / percolation into an aquifer—which may be natural or induced. The process of replenishment of groundwater, by lateral, vertical (or combined) water transfer.

recharge boundary A river or lake which is actively recharging an aquifer, and which is therefore effectively at a fixed hydraulic potential (the water surface potential) along the surface water / groundwater interface.

recharge dam Usually a low-gravity dam designed not to store surface water, but to impound the surface water (or at least some of it) long enough for it to infiltrate into an aquifer. The dam itself may be deliberately 'leaky'.

recharge well = **injection well**. A point source of recharge to an aquifer, which may be either pumped or aquifer fed. The technology of artificial recharge by injection wells is of growing importance.

recharge basin A basin excavated to a horizon which affords highly permeable continuity with an underlying aquifer, and into which excess storm-water, stream-water *etc*, can be diverted for aquifer recharge. In some cases a stream bed itself can be engineered into a cascade of recharge basins.

recovery 'Recovery' has at least 4 specific meanings in hydrology:

Dictionary of Hydrology and Water Resources

(1) That fraction (%) of feed water that is converted to fresh water in a desalination plant.
(2) The decelerating rise in DWL towards its original SWL after a pump has been switched off.
(3) The latter part of a **SAG curve** *(q.v.)* in which aeration brings the DO back to its original unpolluted concentration.
(4) In a hydrocompacted aquifer and/or bounding aquifuge that has undergone de-pressurization, the recovery is the *elastic* expansion that occurs by re-pressurization.

recycling (Of wastewater). The second most important process in water conservation (the first is increasing irrigation efficiency). Wastewater recycling is seldom utilized to anything like its optimum extent. The exception is in Israel, which recycles 70% of its wastewater. Compare Singapore 15%, Spain 12%, and the USA, 0.1%. Recycled water doesn't have to be treated to potable standards. Even minimal treatment may be sufficient for beneficial re-use in irrigation.

redox The redox potential, E_h, is an index of reducing - oxidizing conditions, and is measured in volts or millivolts with respect to a standard hydrogen half-cell potential. Most natural waters are oxidizing (positive) in the range 0 to 0.9 V. Some groundwaters, and hypolimnic, polluted or stagnant surface waters may be slightly to highly reducing (0 to -0.6 V). Reduced surface waters are regarded as ecologically unhealthy, and probably require remediation. Extreme oxidizing conditions can occur under low pH, and extreme reducing conditions can occur at high pH. For example, there are natural hyper-alkaline springs at pH ~12, and E_h of about -0.6 volts in which the water itself breaks down to evolve hydrogen gas. The redox state of a water is only meaningful if all the redox pairs, such as

	nitrate-nitrite, sulphate-sulphide or ferrous-ferric iron, are in thermodynamic equilibrium.
redox electrode	An electrochemical method of measuring the redox potential, Eh. The electrode consists of a reference Pt-half-cell and a silver-halide half-cell. The potential difference between the two is added to the standard hydrogen-Pt potential to give the Eh. The redox electrode is highly sensitive to streaming and spurious oxidation potentials, and may require up to about 80 minutes to reach equilibrium—so don't rush the measurement!
reducer	A borehole connection between larger and smaller diameters.
reef limestone	Reefs, especially those of Quaternary and Tertiary age, often have exceptionally high transmissivities. Also, if the reef is large enough, it may have very high groundwater storage potential. Reefs often make good drilling targets for water supplies, but coastal reefs are probe to saline intrusion.

reference evapotranspiration 'ET_0' *See* **evapotranspiration**.

reflection	(1) The fraction of incident radiation, R, that is reflected from calm surface water is given by: $$R = \frac{1}{2}\left[\frac{\sin^2(i-r)}{\sin^2(i+r)} + \frac{\tan^2(i-r)}{\tan^2(i+r)}\right], \text{ where } i$$ is the angle of incidence, and r is the angle of refraction (both measured from the vertical). This typically averages about 5% and 10% for summer and winter conditions respectively: more during morning and evening, and less at high latitudes.
	(2) Of a water wave, occurs by impact against a cliff, reef, piling etc.

(3) The notional multiplication of image wells across planes of symmetry (recharge or impervious boundaries) in groundwater hydraulics theory.
(4) Seismic reflection: A geophysical technique which is only of use in very deep aquifers. Most water well exploration is relatively shallow, and hence more suited to geophysical investigation by seismic *refraction.*

reflection peak A small, spurious, high-resistance peak on either side of a highly resistive layer on a borehole log. An artifact of the logging system.

reflux = **backflow**, = **return flow**. The process of fluid recycling in any system, but especially of groundwater return to the sea, as in a tidally isolated lagoon. Reflux processes are often of great geochemical significance.

regelation Melting of ice in response to pressure, with subsequent re-freezing of the moist ice surfaces.

regeneration (1) The recovery of a piezometric surface after pumping, as the DWL asymptotically approaches the SWL.
(2) Acid treatment of an ion exchange resin to restore its former cation adsorption capacity.

regional climatic model Increasingly used and popular downscaled versions of GCMs. Regional climate-change effects are urgently needed for planning purposes, and hence regional models are the subject of intensive development. The problem with down-scaled models generally is that they imply a level of predictive precision that they cannot possibly deliver. As with GCMs, RCMs are much better at predicting temperature changes than they are at predicting changes in rainfall, humidity, winds, soil moisture, crop-water requirements, *etc.* All such models were never intended for hydrologic and water resources applications although, in practice, that is how they have come to be used.

regolith Combined soil and weathered bedrock. There is a rough equilibrium between the rate of weathering and the rate of erosion. Where the rate of erosion is high the regolith may be entirely absent. In slowly eroding wet tropical environments the regolith may be tens or (rarely) hundreds of metres thick. Most typically the regolith is about 3 to 8 metres thick. The regolith tends to be more porous than the underlying bedrock, but whether it constitutes an aquifer is largely dependent upon the clay content and distribution.

regulation . . . of a river. Human intervention in the flow regime of a river, by means of channel improvement, interbasin transfer, dams, weirs, locks, *etc.* Partial or total regulation is required to achieve: flood attenuation, maintenance of minimum pool levels and water transport, water resource optimization, and water quality control.

rejecting aquifer An aquifer with no open discharge zone, in which case there can be no groundwater throughflow, and once saturated, there can be no further infiltration.

rejuvenated stream A partial or entire stream whose erosive power and bedload have been increased either by tectonic uplift, or by a fall in the base-level of erosion.

relative density = Specific gravity. The ratio of the mass of a substance to the mass of the same volume of water at 0°C.

relative error The absolute error ÷ true value, usually expressed in %.

relative humidity $= e_d/e_a \cdot 100\%$, a ratio to express the nearness of moist air to saturation, where e_d is the water vapour pressure, and e_a is the saturated vapour pressure at the same temperature.

relative roughness = relative smoothness (1) A parameter much used in pipeflow calculations. It is the ratio, $\frac{\varepsilon}{D}$, where ε a characteristic of the pipe material being used, and D is the inside diameter.
(2) ...*of a stream bed*... The ratio of hydraulic mean radius to particle diameter, D_{84}, of the streambed sediment. (D_{84} is the particle diameter that is $\geq 84\%$ of the streambed particle diameters).

relative vapour pressure $= e/e_0$ where e is the vapour pressure of the soil moisture, and e_0 is the vapour pressure at the surface of open water at the same temperature. Capillary forces cause $e < e_0$.

relative variability A measure of variation in a data set, defined as:

$$\frac{\sum |\mu - x|}{N\mu} \cdot 100\%$$

where μ is the mean, N is the number of data points, and x is a sequential data point.

relaxation A widely used numerical iterative technique in which a grid of interactive data points is operated on by a computer. The initial data set is not at equilibrium, but the grid is held 'constant' whilst each data point is recalculated to be at equilibrium with its adjacent grid points. By scanning all the points in the grid, over several cycles of iteration, the data set (hopefully) converges to an equilibrium state. There are numerous sophisticated variants and 'short cuts'.

relaxation time The interval between a stimulus / disturbance, and the re-attainment of equilibrium. This may be real or simulated by computer. In a transient computer model, each time step must be long enough for the grid to relax (equilibrate). For example, in a two-dimensional transient model of an aquifer's piezometric surface, the relaxation time $\leq \Delta t \leq \dfrac{Sa^2}{4T}$, where Δt is the time step of the model, T is the transmissivity, S is the storativity, and a is the nodal spacing of the grid.

relief intensity The maximum range of elevation within a catchment.

relief ratio Relief intensity ÷ the horizontal distance over which it occurs.

repeat section In nuclear logging (γ, neutron *etc.*), where results are dependent upon the hoist rate, part of the log (usually the lowermost section) is repeated two or more times to optimize the hoist rate. The objective is to find the fastest rate which gives a reasonably reproducible log.

reprofiling Engineered reshaping of a river section, such as to improve bank stability or to provide greater flood storage capacity.

resequent river A specific form of consequent drainage in which a river deviates from conformity with the geological dip, but later conforms to the dip at a lower structural level. This usually happens in response to a lowered base level of erosion.

reservoir stripping The removal of organic topsoil from a reservoir site prior to filling, thereby reducing the BOD and enhancing the subsequent water quality.

residence time   The hydraulic residence time, alias the lake retention time, flushing time or 'water age', is the volume of water (in lake, atmosphere, aquifer, *etc.*) / inflow rate. The sediment, nutrient or solute residence times are longer due to sedimentation, ion exchange, and organic recycling. By this definition, a lake's residence time is shorter than reality, because it assumes efficient mixing (no 'dead spots') which, of course, does not happen if the lake is stratified.

residual chlorine   *See* **chlorine residual**.

residual drawdown   At any given time, t, after the end of a pumping period (especially a pumped aquifer test), the residual drawdown, *r*, is the difference between the actual drawdown and what it would have been had no pumping taken place. The natural piezometric recession is taken into account in estimating the SWL_t.

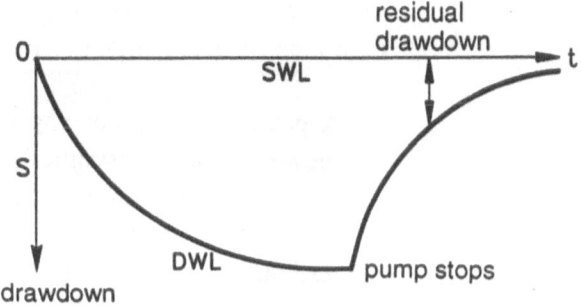

residual oil   = **immobile oil**, = **pellicular** and **molecularly bound oil**. It is mostly associated with the unsaturated zone.

residual pore pressure   When a 'well-behaved' confined aquifer is pumped, it loses pressure almost instantaneously. On the other hand any adjacent clay-rich confining layers (above or below) lose pressure much more slowly due to their very low hydraulic conductivity (inefficient pore drainage). It may take years or even decades for the aquitard's pores to regain pressure equilibrium with the adjacent aquifer. During this time, the high residual pore pressures set up a transient hydraulic gradient, which in turn produces a 'delayed yield' and hydro-compaction of the aquitard.

residual rain   The 'tail end' of a rainstorm in which the precipitation rate < the continuing infiltration rate; the last of the rain which scarcely, if at all, contributes to quickflow.

residual water level   = Recovering water level. The transient piezometric level in a well or piezometer between the end of pumping and the return to natural static water level. The RWL - time relationship for a pumped well can be used to estimate transmissivity, providing the well's effective radius is known.

resistivity   Electrical resistivity, in Ω-m, is a useful parameter both for groundwater prospecting along a horizontal traverse, and for assessing the lithological variation in the vertical profile of a borehole.

respiration   A plant's waste-producing mechanism, essentially the reverse of photosynthesis: $C_3H_6O_3 + 3O_2 \rightarrow 3CO_2 + 3H_2O$ + energy. Unlike transpiration, respiration is a 24-hour process, but the amount of water lost by respiration is trivial compared to that by transpiration.

resurgence   The reappearance of water, especially as karstic springs, but also as upward outflows from engineered conduits such as 'inverted siphons'.

retardation   Ion exchange processes in aquifers can effectively slow the movement of a solute front relative to the rate of groundwater flow. Thus some strongly adsorbed pollutants may be all but immobilized despite relatively fast groundwater velocities. Retardation may be expressed as concentration-time graphs, as shown in the following 'retardation curves' (relative concentration, $\frac{C_t}{C_0}$ vs. time):

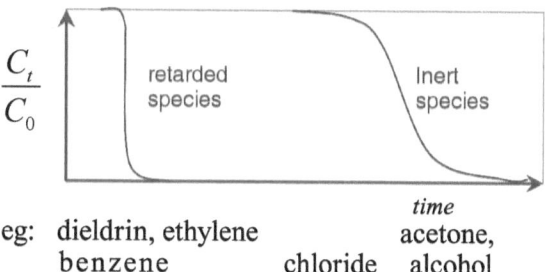

eg: dieldrin, ethylene   acetone,
     benzene   chloride   alcohol

Retardation Factor The ratio of Darcian velocity of groundwater flow to the effective velocity of a solute in the same flowpath. Since some solutes are retarded by adsorption, decay, *etc.*, the retardation factor, 'R'<1.

return period 'T' = Average Recurrence Interval (ARI). This term is gaining in popularity because the synonym of 'return period' was frequently misconstrued as the interval before an event of given magnitude would reappear. No such periodicity is implied, hence the emphasis upon the long-term *average*. There are several definitions, of which the Weibull $T = \frac{N+1}{r}$, and Cunane $T = \frac{N+0.2}{r-0.4}$ are commonly used, where r = rank of events in the record, and N = total number of records).

reversible river Seasonal reverse flow, caused by dominant flow in the main channel backing up into a tributary. Examples are the Blue Nile flood peak backing upstream into the White Nile, or the Mekong backing up into the Tonlé Sap in Cambodia. Not to be confused with 'Reversed river'.

reversed river Rare cases of tectonic uplift causing the permanent reversal of river flow. The classic example is the Ganges which, prior to Himalayan uplift, flowed from east to west, rather than the current west to east.

reverse osmosis A method of desalination in which saline water is highly pressurized against a semi-permeable membrane, such that pure water seeps through to the low pressure side. (Under zero pressure differential, fresh water would diffuse through the membrane in the opposite direction to dilute the saline water—hence the name of the process). Originally, RO was only suited to brackish water, but since the late 1980s, the technology has improved to the stage where it can compete with other methods of sea water desalination.

revetment A pavement of tiles, concrete, riprap, gabions or steel piling installed in the bank of a river, reservoir, lake or coast to moderate or prevent erosion.

rewilding The sensible and desirable process of returning a river to its natural state by reinstating ox-bows, snags, wetlands, etc., often resulting in major reductions in downstream flooding.

Reynolds number $R_e = \dfrac{\rho v y}{\mu} = \dfrac{\rho v r}{\mu} = \dfrac{v y}{V}$, where ρ is the density, v is the velocity, r is the hydraulic mean radius, y is the uniform depth, μ is the coefficient of viscosity, and V is the kinematic viscosity. R_e is a numerical index whose main purpose is to differentiate between laminar flow (R_e < about 500), transitional, and fully turbulent flow (R_e > about 2500).

rheology The study of flow properties, such as thixotropy, deformation, viscosity, bed shear-stress, *etc.*

rhodamine A family of strongly coloured fluorescent organic dyes used in water tracing or dilution gauging experiments, and amenable to measurement at extremely low concentrations. Rhodamine compounds include: Na-fluorescein, Rhodamine B, Rhodamine WT, and Pontacyl pink.

rice Rice cultivation is the single biggest beneficial water consumption on the planet. Most of the world's rice is grown by flood irrigation, even in some desert and semi-desert areas. By utilizing the 'system of rice intensification' it can be grown with about half the amount of water used in flood-irrigation, but SRI is even more labour intensive than traditional methods. Along with cotton and sugar cane, rice has one of the highest water demands of all crops. Ignorance, greed, tradition and inertia often conspire to grow rice under woefully inappropriate conditions. Evaporation from rice paddies is a function of rice variety, soil-moisture retention, growing duration, temperature, humidity and most importantly the duration of sunlight (higher in summer temperate latitudes than in the tropics). In coming decades it will become increasingly essential to greatly improve the irrigation efficiency of most of the world's rice farming areas.

Richards' equation For unsaturated flow: $K_x \frac{\partial^2 H}{\partial x^2} + K_y \frac{\partial^2 H}{\partial y^2} + K_z \frac{\partial^2 H}{\partial z^2} = \frac{\partial \theta}{\partial t}$, where K is the unsaturated hydraulic conductivity in the direction indicated, H is the hydraulic potential, θ is the moisture content, and t is the time.

Richardson number A measure of whether wind shear is sufficient to cause overturn in a stratified lake. R_i is the ratio of turbulent work done by wind shear, to buoyancy of the epilimnion. If it is less than about 0.25, then mixing will occur. R_i is a dimensionless quantity given by:

$$R_i = \frac{g \cdot \partial\rho/\partial z}{\rho \left[\partial V/\partial z\right]^2}$$

where $g \cdot \partial\rho/\partial z$ expresses the buoyancy, and $\partial V/\partial z$ is the vertical velocity gradient caused by 'stirring' due to wind shear.

riffle Shallows in a river, characterized by rapids, turbulence and, perhaps, by standing waves.

rig A drilling rig, and associated pump, drive mechanism *etc.*

rillenkarren Dissolution grooves which develop on steeply sloping limestone surfaces exposed to rainfall or condensation. The width varies from a mm to several cm, and the length may occasionally exceed a metre. Sharp sub-vertical ridges may develop between the grooves.

rime An irregular porous ice layer consisting of scaly or feathery crystals, separated by air. It forms by the impact and sudden freezing of supercooled water droplets.

Rimmar formula One of several empirical methods for estimating the optimum mixing length for use in dilution gauging.

$$L \propto \frac{b^2 v^2}{d^2 s}$$

where b is the stream width, v is the stream velocity, d is the hydraulic mean depth, and s is the stream gradient. A coaxial nomogram relates the variables, as in IASH 78, p 401.

ring infiltrometer Also known as **double-ring infiltrometer**. As the name implies, this consists of two concentric tubes, usually upwards of 15 cm and 30 cm diameter, which are impressed into the soil, for perhaps 5 cm. The water level in the outer section is maintained at constant head, whilst in the center section alternative methodologies can estimate the infiltration rate from either a falling or constant head. With small diameter rings errors of up to ~30% may occur. Naturally, the larger the ring diameters the more accurate the infiltration estimate.

riparian Pertaining to a riverside, such as in river water rights, land use, vegetation, and in the case of international rivers, even countries.

ripe snow Old, coarsely crystalline snow, of specific gravity 0.3 to 0.6, at incipient melting point.

riprap Boulders or rough cut stone used for bank protection works.

riser A vertical water supply pipe, as in the feed to an elevated water tank, or the rising main which feeds water from a submersible pump in a borehole, to the surface reticulation system.

risk The *hydrologic* risk, R, (of exceedence or failure) is the probability that a flood of given average recurrence interval will be exceeded at least once within a period of N consecutive years. It is calculated as $R = 1 - \left(1 - \frac{1}{ARI}\right)^N$.

In recent years some confusion has arisen from two connotations of risk being used in the same project. In the *climate change* context, 'risk' combines the magnitude of a climatic impact with the probability of its occurrence. The *climate change risk* has been defined several different ways, both objectively and subjectively.

river capture = **river piracy**. A common geomorphic feature in which a headwater from one catchment cuts back into the fluvial system of an adjacent catchment, thereby intercepting the drainage and diverting it into the more active channel. Groundwater capture is less obvious, but it is possible for the active river to cut back and capture groundwater from an adjacent catchment, without capturing the surface flow.

RORB A popular and quite successful model of runoff and streamflow routing for estimating flood hydrographs. The model simulates a catchment as a network of non-linear concentrated storages. The package has various routines for simulating urbanization, flood storage *etc.*

rotation The direction of rotation of water draining through a plughole is said to be clockwise in the northern hemisphere, and counter-clockwise in the southern hemisphere, in response to the Coriolis effect. This is demonstrably false nonsense, as pipe friction in the drainpipe is many orders of magnitude greater than the miniscule Coriolis effect.

rotary rig Truck-mounted equipment for direct rotary drilling, including the mast, rotary table, compressor, mud pump, and control station.

rotifer Very common aquatic ciliate metazoans. There are some 2000 species varying in size between about 0.1 and 1.0 mm long.

rotodynamic machine Pumps, turbines or impellors. A class of machines driven by, or which drive, moving water.

roughness coefficient = An index of stream-bed roughness: The 'n' in Manning's equation. It varies with stage. Typical values are:

planed wood	about 0.012
concrete or smooth earth channels	0.016 to 0.022
earth channels with weeds, bends etc.	0.025 to 0.033
sand and gravel channels (deep to shallow)	0.029 to 0.058
rubble or rock channels (smooth to rough)	0.025 to 0.050
sluggish weedy reaches with deep pools	0.050 to 0.080
streams cascading over boulders	0.030 to 0.070
overbank flow onto low grass or sparse vegetation	0.025 to 0.035
overbank flow onto long grass to dense woodland	0.040 to 0.200

There are several 'fiddle factors' available to correct 'n' in respect of cross-section irregularity, vegetation, obstructions, sinuosity, etc. *See also* **Strickland formula**.

rough pipe   A pipe whose raised imperfections on the inside surface are sufficiently large to penetrate the viscous sub-layer, and hence to retard the flow.

routing   A numerical technique, usually computerized, of calculating the timing and changing shape of a flood wave as it moves downstream. The method is applied to longer rivers as a means of flood warning and/or planning for river control. Strictly, river routing requires numerical solutions to the complex Saint Venant equations, which in turn requires a lot of expensive data. In practice computer models simplify much of the data requirements to comprise:

i) Historic discharge measurements,
ii) The channel geometry, including storage volume along each reach of a river,
iii) the river surface slope, and
iv) the roughness coefficient.

The basic routing equation for each reach uses the time incremental expression: *'inflow=outflow + rate of change of storage'*.

rule curve Also known as **critical rule curve**. The ideal schedule of reservoir operation; the optimized graph of water level *vs.* time of year, which reservoir managers use as a target for controlling the available water resource. No rule curve can be regarded as uniquely correct, and there may be multiple rule curves corresponding to differing drought scenarios. The curve varies with available runoff, storm and drought frequency, climatic drift, and water usage. In the case of shared international boundaries there may also be international treaty constraints upon the rule curve. It is common for sector conflicts to occur, with irrigation, flood control and hydro-power stakeholders all arguing to apply different rule curves to the same reservoir storage. In such cases MCA and basin hydrology software may be needed to reach a workable compromise.

rundkarren A karst feature of rounded dissolution grooves, without sharp edges, that develops at a soil-limestone interface.

runnel A rivulet.

runoff = Streamflow = catchment yield = discharge, (units of cumecs or $l.s^{-1}$, although some unscientific authorities persist in using 'acre-feet per day'). Prediction of runoff from rainfall and catchment characteristics is one of the major tasks of the hydrologist.

runoff coefficient The fraction of precipitation which results in quickflow. It varies with catchment characteristics, seasonality, total precipitation and antecedent conditions. In moist mid-latitudes, the overall runoff coefficient is typically about 70%, but the range is very variable.

Ryznar stability index Defined as $I = S - C - pH$, where S and C are determined from the total dissolved solids, methyl orange alkalinity, and the Calcium concentrations (mg.l^{-1}). $I > 7$ indicates corrosive water. $I < 7$ indicates likely carbonate encrustation.

S

sabkha (Arabic) An actively precipitating evaporitic environment, especially intertidal shallows bordering the Arabian Gulf. More generally, sabkhas include any inland playa or evaporating groundwater discharge zone.

sacrificial anode An electropositive metal connected to a relatively electro-negative metal, thus setting up a galvanic couple which oxidizes the sacrificial anode (electropositive) whilst protecting the cathode. For example, a steel borehole in moist soil could be protected from rusting by connecting it to a sacrificial anode, such as a piece of scrap magnesium, which is also buried in the soil.

safe yield = **sustainable yield**; the maximum water extraction that can be periodically replenished. This may refer to a well, wellfield, aquifer or entire river basin. It is commonly equated to the rate of 'aquifer recharge', but the 'safe yield' is not a fixed value. Rather, it varies somewhat according to technology, economics, time-scale of extraction, and above all, with the wisdom of resource managers! In this era of climate change it is prudent to re-examine old estimates of the 'safe yield'.

safety factor (f). An arbitrary or legally required multiplier applied to a design maximum stress / capacity / yield, *etc*, to which a structure is designed. For example, if a reservoir is expected to just withstand a maximum hydrostatic pressure of P, then it is designed to withstand $f.P$.

Saffir-Simpson scale The most widely used classification of hurricanes and tropical storms, using sustained wind speeds as the indicator:

Category	Wind Speed in km.hr^{-1}
Tropical depression	<62
Tropical storm	63 to 118
Hurricane 1	119 to 153
Hurricane 2	154 to 177
Hurricane 3	178 to 208
Hurricane 4	209 to 251
Hurricane 5	>252

sag curve A graph of % saturated dissolved oxygen *vs.* distance - time. This is characteristic of surface water downstream of a point discharge of an organic rich effluent (such as a sewage outfall). It is modelled by an equation of the form;

$$D_t = \frac{K_1.L_0}{K_2 - K_1}\left(e^{-K_2 t} - e^{-K_1 t}\right) + D_0 e^{-K_2 t}$$ where K_1 and K_2 are reaction rate constants for deoxygenation and re-aeration respectively, D_0 and D_t are the dissolved

oxygen deficits (initial and after time t), and L_0 is the initial biological oxygen deficiency.

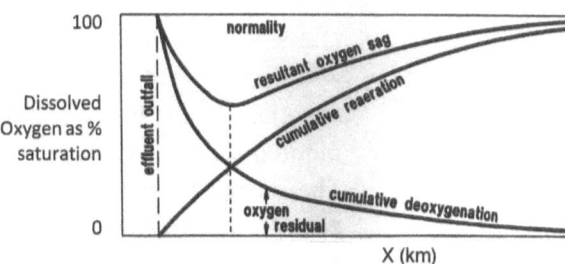

Saint Venant Equations Unsteady one-dimensional surface flow, such as a flash-flood or dam burst, is described by the two Saint Venant equations. The first is a conservation of mass equation whilst the second is a conservation of momentum equation, respectively:

$$\frac{\partial Q}{\partial x} = \frac{\partial}{\partial t}(A + A_0) - q = 0 \ldots\ldots\ldots\ldots\ldots ①, \text{ and}$$

$$\frac{\partial Q}{\partial t} = \frac{\partial}{\partial x}\left(\frac{Q^2}{A}\right) + gA\left(\frac{\partial x}{\partial z} + S_f\right) = 0 \ldots\ldots\ldots ②,$$

where Q is the mean discharge in the direction of flow, t is the time, x is the distance in the direction of flow, z is the depth of flow, g is the gravitational acceleration, S_f is the friction gradient, A is the *active* cross-sectional area of flow, A_0 is the cross-sectional area of *off-channel storage* (such as overbank flow, or underflow in the bed sediments), and q is the net inflow (+) or outflow (-) per unit distance of flow. There are several variants. For example the second term of equation ② may be modified by α, (the velocity head coefficient) if α is not ≈ 1.0; whilst the third term may include additional 'S' factors to model channel expansion or contraction, or any non-Newtonian fluid behaviour (as in mudflows). Also a fourth term may be added to allow for the momentum of lateral inflows. In equation ①, the second term may be modified by a 'sinuosity coefficient' to account for helicoidal flow. In both ① and ② the $\frac{\partial}{\partial x}$ ... term is the 'unsteadiness', without which the equations revert to the simpler and generally more familiar steady-state equations of flow.

S.V.Es continued The second Saint Venant equation is very difficult to solve analytically, but software is available to estimate numerical solutions.

salina = **salar**, = **schott**, = **playa**. An ephemeral salt lake or its desiccated equivalent.

saline interface The notional interface between dense sea water and relatively light freshwater in a coastal aquifer (and to a lesser extent in inland stratified aquifers). Hydrostatic and dynamic models incur only minor errors by assuming that the interface is sharp although, in reality, it is more like a zone of diffusion.

saline intrusion = **saline invasion** ≈ **sea water encroachment**. The induced migration of saline water into aquifers, generally due to mis-management or overpumping of aquifers in coastal situations. The accelerating rise of sea level, due to global warming, bodes ill for most of the world's coastal aquifers.

salinisation = **salinization** = **salination**. Destruction of agricultural viability by increasing salinity. The three forms are: dryland, irrigation and coastal salinisation. This is one of the four most serious global environmental degradation processes, together with desertification, soil erosion and urbanization (loss of habitat).

salinity

The transition between fresh and saline water is arbitrarily set at 3‰ (or 5‰ according to some authorities). This compares with sea water: 35‰ and saturated NaCl solution: about 350‰. In the case of sea water, since the Cl⁻ ion is well mixed, conservative, and approximately constant, salinity = 1.8065 x chlorinity. Alternatively, the salinity may be given in *total dissolved solids*, as shown below:

		fresh	brackish	saline	hypersaline
T.D.S.	from	0	10^3	10^4	$>10^5$
mg.l⁻¹	to	10^3	10^4	10^5	-

saltation

Strictly, particle transport by saltation is an aeolean mechanism, but saltation-like motion of stream-bed particles does occur, in which one particle is dislodged by the impact of another particle, and briefly advected. The overall flux by saltation may be uniform, but individual particle movement is strongly episodic.

salt balance

Usually calculated as the chloride mass-balance between inflow and outflow to / from a catchment. In a catchment of low salt storage and relatively short chloride residence times, the salt balance is one of the most useful quantitative tools for understanding the water balance.

Salmonella

A family of pathogenic bacteria including *S. typhi*, and *S. paratyphi-A*, which gives rise to typhoid and paratyphoid fever. Other species can cause moderate to severe gastro-enteritis. Although Salmonella is mostly associated with food poisoning, it can also be transmitted through water.

Salvinia

(*S.auriculata*). A troublesome free-floating tropical hydrophyte, originally from Brazil, but now worldwide. It spreads rapidly. For example, in only a few years, 500 km² of Lake Kariba was covered in *Salvinia*. Consequences include the interception of light, water de-oxygenation, and the death of phytoplankton and fish. Total eradication is extremely difficult, but *Salvinia* can be controlled by herbicides, biological or mechanical methods.

sand Rock particles in the size range 63 μm to 2.0 mm. 'Sand' does not have any compositional connotations.

sanding up Sedimentation in the sump of a borehole. To compensate for this, bores often extend for several metres below the active screen.

sand trap/gate = **sediment trap**. A gated opening in a section of channel, with a large cross-sectional area. Sand traps are set up in experimental streams or irrigation channels to allow the medium-to-coarse sediment to settle out.

sap flow Water movement inside a tree, from root to leaf, as distinct from transpiration, which is the movement of water from leaf to air. It is generally measured in the lower stem/trunk by a stem **psychrometer** *(q.v.)*. It is called sap flow rather than water flow because the water carries with it an abundance of plant nutrients. Sap flow is driven by photosynthesis, so its rate is roughly proportional to the leaf area and incident sunlight, but it is also affected by the availability of soil moisture and the tree's health.

saprobicity A measure of biodegradable organic matter in a stream.

sapropel Gelatinous black sludge characteristic of anaerobic lake beds.

saqia = **sakiya**. An ancient mechanism of lifting water for irrigation. It consists of two enmeshed wheels with axes at right angles. The horizontal wheel is rotated by an ox, mule etc, and is used to drive a vertical wheel with numerous buckets. The lift is up to six metres. Two oxen can lift up to about 300 m^3.hr^{-1}. Saqias are still widely used in parts of India and the Nile Valley.

saturated zone This is *not* defined as 'all parts of an aquifer in which the pores are 100% filled with water'. Rather, the s.z. is 'all parts of the aquifer in which the hydrostatic pressure is > atmospheric pressure'. In practise this is almost the same, but the capillary fringe is not included in the s.z., whilst there may be bubbles of air entrapped in the s.z., especially with a rising water table.

saturation deficit
(1) At a given temperature and pressure, the additional moisture required to bring *air* to saturation, *i.e.*: $e_a - e_d$.
(2) The water depth, in mm, required to bring a soil surface to saturation (including maximum wetting of depression storage). However, saturation of the entire section to the water table is not necessarily implied. Since the wetting front is mobile, the *soil* saturation deficit is variable, and inadequately defined.
(3) In a solution, the saturation deficit is $\dfrac{C_s - C}{C_s}.100\%$, where C is the solute concentration and C_s is the saturated concentration of the solute at the same temperature.

saturation index
(1) $SI - \log\left(\dfrac{IAP}{K_s}\right)$, where the Ion Activity Product is a function of the actual solute

concentration, and K_s (or KT in alternative notation) is the equilibrium solubility constant. The three possibilities are: undersaturation if S.I. < 0, exact saturation if S.I. = 0, and supersaturation (= oversaturation) if S.I. > 0. *Important*: The saturation index is only meaningful in the case of equilibrium thermodynamic conditions.

(2) The **Langelier Index** *(q.v.)* is also sometimes referred to as the saturation index. This practice is confusing, and should be dropped.

savannah — Or savanna. Tropical grassland, often with scattered low phreatophytic trees, such as Acacias.

SCADA — 'Supervisory Control and Data Acquisition'. Centralized industrial control systems, such as required for large pumping, irrigation, drainage, pipe-line and water distribution systems, are all heavily reliant upon SCADA to quantify and manage the complex interactions, to cut wastage and optimize water supply where and when it is needed. The SCADA probably uses a programmable logic controller, remote data monitoring such as water level sensors, remote control such as irrigation gates, automatic safety cut-outs for leakage and unexpected events, a system database, security protection, and of course, some form of human oversight based in/around a control centre.

scald — An erosive depression in which topsoil is removed, and a crust develops at the previous topsoil - subsoil boundary. Most scalds are associated with dryland salinity, but sodic and non-saline scalds also occur.

SCC — Surface Contact Cover. Any soil surface cover, such as a mulch, leaf litter or low vegetation, which absorbs raindrop impact and erosion, and which inhibits runoff.

schistosomiasis = **bilharzia**. Various species of water snail are the intermediate hosts for this infectious parasitic fluke. Over two million people are infected throughout the tropics despite great efforts to eliminate the snail vector. Chemical control of the snail has largely succeeded in China and elsewhere, but increasing irrigation throughout the tropics and subtropics has also extended the range of the disease.

Schlumberger array In shallow resistivity surveying, a symmetrical configuration of four electrodes in which the inner (potential) electrodes are much closer together than the outer (current) electrodes. Schlumberger, Wenner and 'offset-Wenner' are the most commonly used electrode configurations.

Schoklitsch equation In 1926 Schoklitsch developed an equation for sediment load in a river, as an improvement upon the earlier 'Du Boys equation' (1879). Both equations employed the concept of 'bed shear stress'. It is an empirical expression for the critical discharge, q_c, per unit width of channel.

$$q_c = 0.26 \frac{(\gamma_s - \gamma_w)^{1.67} . d_6^{1.5}}{\gamma_w . s^{1.167}},$$ where $\gamma = \rho.g$, d_6 is the sixth decile particle size, and S is the streambed gradient. This equation works reasonably well for coarse high-energy sediments. The Schoklitsch equation is widely used, but like all such empirical equations, it is a simplification of the complex physics of sediment transport, and therefore should be verified by physical observations.

Schoeller diagram A visually chaotic and clumsy nomographic diagram of a water chemistry; best consigned to the trash-can of history.

Scoby's formula ... for flow in pipes: $q = C \dfrac{\bar{v}^{1.9}}{d^{1.1}}$, where C is a friction constant, $\bar{v}$ is the velocity and d is the pipe's internal diameter. Compare **Darcy-Weisbach**, **Hazen-Williams** and **Manning's** equations.

Scotch mist = **mizzle**. Precipitation in the form of ground level advection of cloud droplets.

scour Stream bed erosion by fast flowing water. The processes are surprisingly complex and difficult to simulate. Contributing variables include density, particle size and distribution, armouring, sediment supply, velocity and turbulence, and the opposing processes of erosion and deposition. Most rivers are in a state of quasi-equilibrium between particle mobilization and deposition (except during floods). Thus, if the suspended sediment concentration is reduced, as downstream from a dam, then the scour rate is dramatically increased until a new equilibrium is found.

scour equations A common hydrologic problem is how to estimate the depth of scour around a pier, such as for bridge foundation design. There are many scour equations, all of which incur some degree of uncertainty due to variable particle size around the pier, unsteady flow, and the degree of armouring. One well-validated equation, by Hafez (2004), is:

$$\left(\frac{D_s}{H}\right)^3 = \left(\frac{3\tan\phi}{(S_g-1)\cdot(1-\theta)}\right) \cdot \frac{\eta^2 v_x^2}{gH\left(1-\dfrac{b}{B}\right)^2} \cdot \left(1 + \frac{D_s}{H}\right)$$

where the parameters are as below and in the diagram.

This is a cubic equation, and hence has to be solved iteratively. River long section: depth H, depth of scour D_s, mean approach velocity v_x, pier width b, where B is the channel width, or pier to pier centerline width. ϕ is the bed-material angle of repose, θ is the bed-material porosity, S_g is the specific gravity of the sediment, and η is a transfer coefficient of the horizontal momentum into vertical momentum in the downward direction $\eta V_x = V_z$. The latter is determined empirically or by numerical modeling.

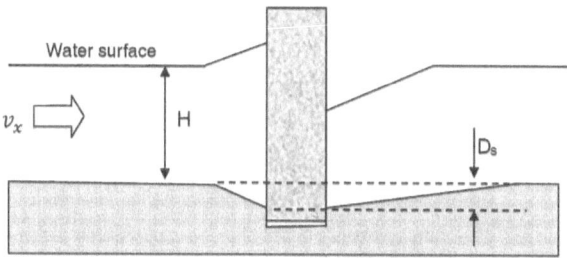

Screen That part of a borehole consisting of a slotted tube or wire-wound tubular frame which permits the inflow of groundwater to the well whilst maintaining its structural integrity. The 'slot size' depends upon the particle size of the aquifer to prevent clogging. Much of the skill in well logging is devoted to setting the screen at the most appropriate depth.

screw pump *See* **Archimedian screw.**

S-curve A graph of the 'cumulative time unit hydrograph'. This hydrograph results from indefinite continuous rainfall of intensity 1/T mm.h^{-1}. It is used to calculate unit hydrographs of differing duration.

sealed source Any source of radiation that is completely enclosed in its container. Therefore, radiation can escape, but the source material cannot.

sealing Soil surfaces subjected to continued infiltration tend to 'self-seal'. The soil physics involved is complex, and is still not fully researched. Sealing in unlined irrigation ditches is of economic importance, and may be mediated by microbiota.

Secchi disk A disk marked in black and white quadrants, which is lowered into turbid water until it can barely be seen. The resulting depth is a measure of the turbidity.

secondary baseflow *See* **interflow**.

secondary flow Components of flow perpendicular to the main direction of flow. Examples include helicoidal flow, or contraction immediately upstream of non-suppressed weirs.

secondary porosity *(As opposed to primary or intergranular porosity).* Any porosity that develops after deposition, especially due to tectonic or dissolution processes. Most secondary porosity is strongly anisotropic such as faults, joints, bedding planes, shrinkage voids, and karstic development.

secondary treatment A range of mostly biological processes, by which filtered sewage or other wastewater is cleansed. Secondary treatment should remove most, if not all, of the organics (and hence the BOD), together with any residual particulate matter. For high quality re-used water, tertiary treatment may still be required.

secondary salinisation Human induced salinisation.

section control = **complete control**. Streamflow conditions in which stage is uniquely related to discharge. For example, immediately up-stream of a transition from subcritical to supercritical flow.

sediment delivery ratio The ratio of sediment yield to total erosion in a catchment; the proportion of eroded material removed from a catchment by fluvial transport.

sediment discharge curve = **sediment runoff curve**. A graph of stage *vs.* sediment load. The graph is different for rising and falling stages.

sediment hydrograph A graph of fluvial sedimentary flux (both bed load and suspended load) *vs.* time.

sediment load The weight percent of a stream which is transported sediment. It normally varies up to about 1%, but rivers in loess or glacial areas may have exceptional sediment loads of up to about 5%. The sediment load may be regarded as the sum of wash-load, suspended-load and bed-load, although the distinction between bed-load and suspended load may, at times, be unclear. In the medium-term climate-change is likely to greatly increase the sediment load in those rivers which are fed by glaciers and/or snowmelt, or in those catchments which were glacial during the Pleistocene. In the very long-term (millennia) their sediment load is projected to decline again as the sediment source areas become relatively depleted. In many rivers most of the annual sediment load may be transported within the few hours of peak flow, when velocities are much higher than normal.

sedimentation rate When designing a reservoir or dam it is essential to know the sedimentation rate, and hence the effective life of the investment. An initial estimate of the sedimentation rate can be made from empirical equations. In the long term the *only* reliable estimate is evidence-based, using 'before and after' bathymetry. This can only be achieved if the reservoir is accurately surveyed during or soon after construction. Note that sedimentation is most intense at the inflow end of the reservoir.

sediment sump = **rat hole** = **tail pipe**. Casing below the screen used as a fine-grained sediment trap.

sediment yield ... *of a catchment* ... The bed load and suspended load of a river, usually quoted on a mean annual basis.

seeding *See* **cloud seeding.**

seepage Diffused groundwater emergence at the surface, as opposed to a spring (point emergence).

seepage face The areal extent of Darcian flow exposed at the surface or in a mining / foundation operation.

seepage velocity = **Darcian velocity** ÷ porosity. The true mean rate of ground-water flow around particles in the aquifer. Seepage velocity is always > the Darcian velocity.

seep zone Saturated soil near the base of a slope, where interflow and groundwater flow emerge to become surface flow.

segregated ice The migration of pore water to a freezing front, resulting in frozen soil - ice with a planar texture.

seiche = **gravity wave**. A feature of lakes in which a hydraulic gradient set up by wind friction subsequently oscillates with a natural period, t, of: $t = \dfrac{2l}{n\sqrt{gd}}$, where n is an integer, normally 1, but occasionally 2 or 3 depending upon the number of nodal axes. l is the maximum fetch of the lake, and d the mean depth. The period typically varies from a minute to several hours. For a lake of several kilometres in length, a typical seiching amplitude would be about 10 cm, but seiching of over a metre in amplitude can occur on large lakes, such as observed events in Lakes Geneva and Michigan (maximum 1 and 3 metres respectively). Seiching periodicity may be complicated by the wind speed, lake stratification and sub-basins of differing depth. *cf.* **period of oscillation**. Less frequently, seiching can also be caused by seismic events.

seismicity An effect of large dams and reservoir construction is the increased risk of seismicity. The chain of causality is a rise in lake level *increasing* the hydrostatic pressure beneath and around the reservoir. This *decreases* the inter-granular pressure of any clastic rocks, and hence effectively weakens the rock. Simultaneously, differential rock pressure, caused by the weight of the impounded water, causes micro-seismicity. In the case of pre-existing faults and fractures, or of rapid drawdown, severe storms, *etc.*, the compounded stresses can give rise to massive slope collapse. The classic case is the Viant dam disaster of 1963, where a major storm and minor earthquake caused 115 Mm^3 of rock to collapse into the reservoir, causing a flood wave which killed 3000 people.

selectivity series Different ions have varying degrees of affinity for adsorption at an exchange site (due to differing charge densities). Ranking of this affinity is somewhat variable, but may be approximated by the selectivity series: $Ba^{2+}>Sr^{2+}>Ca^{2+}>Mg^{2+}$ (divalent), and $Cs^+>Rb^+>NH_4^+>K^+>H^+>Na^+>Li^+$ (monovalent).

selenium (Se) A teratogenic element with a chemistry somewhat similar to sulphur. Large concentrations of Se are immobilised in reducing environments such as coal, sulphides and marine shales, but when oxidised, much more soluble species, such as SeO_3^{2-}, and

HSeO$_3^-$ (selenite and biselenite) are mobilised. Several major aquifers in the western USA and elsewhere are currently suffering increasing Se pollution due to large drawdowns which create new oxidising unsaturated zones from which biselenite is reached.

self potential = **spontaneous potential**. A geophysical technique in which natural variations in electrical potential are measured between surface points or between the surface and successive depths of a borehole (the bore being filled with a conducting drilling fluid). In boreholes the SP log is normally run in conjunction with the resistivity log.

semi-aridity There are so many connotations in the literature that no definitive entry is possible. Broadly, any climate with a perennial monthly moisture deficit of between -20 and -40 mm is considered semi-arid.

semi-confined An aquifer is semi-confined if pressure release (by pumping) induces significant leakage from one or both bounding aquitards, *and* if flow in the aquifer is essentially horizontal whilst leakage into the aquifer is essentially vertical. If leakage through an aquitard also has a significant horizontal component, then the aquifer is semi-unconfined.

semi-variogram The basic tool of geostatistics. A graph of variance *vs.* inter-sample distance.

Senescence Literally, the ageing of plants; the temperature and enzyme-driven process of death or dormancy of leaves. In perennial plants senescence is the seasonal shut-down of sap-flow and photosynthesis. The most visible aspect is chlorophyll degradation, during which most of the leaf nutrients are resorbed into the stem tissue.

Dictionary of Hydrology and Water Resources

senility The final stages of a mature river, characterized by sluggish flow, barely above the base level of erosion, and meandering through a wide, very low-relief floodplain.

sensitivity
(1) ... *of a measuring instrument* ... The rate of change of measurement with concentration: $\dfrac{dx}{dC}$.

(2) ... *of a soil*. ... The ratio of soil strength (shear strength) in the natural undisturbed (field) state, to its strength under disturbed (laboratory) conditions *at the same moisture content*. Pre-consolidation reduces the sensitivity of a clay, whilst a high water content and / or a liquidity index of >1 increases the sensitivity. The classification is:

insensitive clays	<1
low-sensitivity clays	1-2
medium-sensitivity clays	2-4
sensitive clays	4-8
extra-sensitive clays	>8
quickclays	>16

Most soil sensitivities are between 0.9 and 1.0. *cf.* **quickclays**.

separation
(1) Separation occurs at a discontinuity of streamlines, or at a discontinuity between laminar and turbulent flow.
(2) Bubble separation, or 'cavitation', when the velocity head falls below the vapour pressure.
(3) A curling trace of vortices caused by separation of a boundary layer from a solid surface.

service level A subjective measure of convenience and accessibility of sanitation and water supply; especially used in the planning of resources in developing countries.

set casing To install well casing which, depending upon the detailed geology may also involve setting spacers, gravel-pack and/or 'telescoping'. If there are environmental issues, setting the casing may also have to take account of local regulations.

set up Wind set up on a lake is the process by which wind shear pushes water to the downwind end of a lake, thereby creating a non-horizontal surface. Release of the wind shear results in a damped harmonic oscillation of the lake levels from one end to the other.

sewage Liquid and solid excreta. The main source of urban pollution of surface and groundwaters. The pollutants are mainly micro-organisms (innocuous and pathogenic), phosphates, nitrates, and organic nutrients.

sewerage The piped system by which sewage is transported to a treatment or disposal site.

shaduf (Arabic). A cantilever-bailed well (bucket and counterweight), suitable for small water supplies from a shallow water table.

shakehole A doline which displays collapse features, characteristic of very advanced karstic development.

shale One of the most common types of sedimentary rock, primarily consisting of clay minerals. Prolonged compression has squeezed out most, if not all, of the water, and has imparted a distinct planar fabric to the rock texture. Shale is invariably an aquiclude or aquitard.

shale line = **shale base line** = **clay line**. The reference line on an S.P. borehole log, defined by the 100% clay signal. In theory, it should be a vertical line, but it

Dictionary of Hydrology and Water Resources

may shift slightly due to stray ground, salinity and system potentials.

shale shaker Equipment on a rotary drilling rig for separating the cuttings from the drilling fluid.

shallow karst A karstic aquifer which is entirely higher than the outflow point. Note that 'shallow karst' is unrelated to either aquifer thickness or depth to the phreatic surface.

shape file For a GIS to operate, data are required to be linked spatially. This is achieved through 'shape files' in which any parameter—well location, topography, water level *etc.*, is linked to an array of corresponding geo-referenced coordinates.

shear zone An imbricate, or brecciated concentration of fracture planes, often forming a locally high-yielding planar aquifer.

sheet flow A thin sheet of water typically in the order of a mm in depth, flowing at the land surface, especially over bare rocks. It occurs only when the rainfall intensity > infiltration rate, and when all surface storage has been filled. Sheet flow conditions approximate to Hortonian conditions types 2 and 3.

sheet piling Interlocking strips of steel driven into unconsolidated ground to form either an impervious screen, or to prevent slumping of surface / near-surface sediments. They may be temporary, as in foundation, quarrying or land-fill operations, or permanent.

shields . . . such as Nipher, Tretyakov and Alter shields. Various devices designed to deflect or absorb micro-aerodynamic turbulence around the rim of a precipitation gauge, thus giving more accurate measurement.

Shields diagram — A log-log plot of dimensionless shear stress, Ψ, vs. the boundary Reynolds number. The upper field of the diagram indicates particle erosion and transport, whilst the lower indicates stream bed sedimentation. The diagram is best suited to fine or medium particle sizes. *NB:* the boundary Reynolds number is: $\dfrac{\overline{V}.d}{v}$, where $\overline{V}$ is the fluid velocity, d is the mean particle diameter, and v is the kinematic viscosity. $\Psi = \dfrac{\tau}{(\rho_s - \rho_w).dg}$, where τ is the bed shear stress, and ρ is the density (of sediment or water). *See also* **Hjulström diagram**.

Shigella — Shigella is the most common aquatic bacterial pathogen, and can infect from very low levels of contamination. Fortunately, it survives poorly in water; typically for 0.5 to 4 hours in warm water. *S. disenteriae* and *S. flexneri* are the sources of bacillic dysentery. Other strains cause mild diarrhoea.

shirkat — A cooperative farm in the former Soviet Union. It is the successor of former 'kolkhoz'. Since the fall of the Soviet Union, most shirkat irrigation infrastructure is in a poor state of repair, leading to low irrigation efficiency.

shoal — A shallow submerged sandbank in a river or lake (or sea).

short normal — A 16 inch resistivity probe used in borehole logging, which gives good spatial resolution, but poor accuracy. *cf.* **long normal**.

shoulder ditch	A ditch dug along the top edge of a slope to prevent erosion.
shrinkage limit	The water required to completely saturate the pores of an oven-dried soil or sediment, expressed as weight % of the dry sample. The weight equivalent of effective porosity (vol/vol).
side contraction	The reduction in width of a stream as it flows over a sharp-crested weir. It is mainly caused by components of flow perpendicular to the main flow direction. *See* **suppression**.
sieve analysis	≈ **particle size analysis**. A graph of cumulative percentage retained (in a nest of sieves) *vs* particle size (sieve mesh size), for any uncemented sediment. The % passing is plotted against grain size represented by each sieve to yield as 'S' shaped graph. A steep curve indicates a well-sorted sediment, whereas a shallow curve is obtained from a heterogeneous sediment. Such analyses are needed for sediment transport analysis, and for the estimation of optimum screen size and gravel pack size in a borehole.

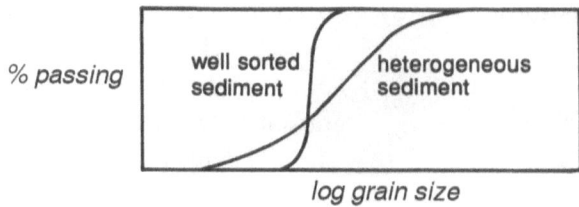

Sievert	(Sv). The unit of effective radiation dose = absolute dose x quality factor.
silcrete	A hard indurated surficial deposit found in both humid and arid environments, and consisting of > 85% silica. The controlling mechanisms of silcrete

silica
formation are not yet fully understood. Silcretes severely inhibit infiltration.

A dissolved constituent of nearly all natural waters. Natural concentrations typically range between about 5 and 80 mg.l^{-1}. Silica hydrates to undissociated silicic acid (H_4SiO_4) except under alkaline conditions, when the solubility greatly increases. Hence, silica is one of the few solutes that does not contribute to the ion-balance. Silica deficiency can result in skeletal and joint problems, whilst high doses in drinking water are harmless, perhaps even beneficial. However, some of the health benefits claimed on the internet are preposterous.

siltation
The clogging of a lake or reservoir by settlement of suspended solids, typically, but not necessarily, of clay grade (2 to 63 mm particle size diameter). Siltation in a reservoir is greatest in inflow deltas (unless the channel is narrow) and in reservoir embayments or tributaries. It is least in the main channel. Thus siltation decreases live storage early rather than later in the lifetime of a reservoir.

similitude
For any physical model to be valid it is seldom enough for the model and real case to be geometrically similar. For kinematic similitude the velocities and velocity gradients of model and real case must have the same ratio. For dynamic similitude, forces, momentum and viscosity must also be similar. Dimensional analysis is usually required to ensure similitude in a model.

Simpsons rule
A method of calculating cross-sectional areas where n is even and $h = \dfrac{(b-a)}{n}, A = \int_a^b f(x).dx$, as approximated by

$$A \approx \frac{h}{3}\{(y_0+y_n)+4(y_1+y_3+..+y_{n-1})+2(y_2+y_4+..+y_{n-2})\}$$

simulation game   A real-time training exercise, and not as frivolous as the term suggests. An essential part of preparedness for major floods, especially for personnel who have never experienced a long recurrence interval event.

sink   A point or dispersed loss of water from a catchment or system. It is modelled by a fall in head (although in reality, it is extracted upwards).

sinkhole   = swallet = sink. *See* **swallow hole**.

sinker   (1) Dense liquid pollutant, such as a brine or DNAPL, that settles to the base of an aquifer.
(2) A heavy streamlined weight on the end of a cable, used to position a current meter in fast flowing water.

sinuosity   A parameter used in river geomorphology and some stream competence (entrainment) calculations. It is also known as Tortuosity, = l/L where l is the arc wavelength and L is the axial wavelength. Alternatively, sinuosity is the ratio of stream length to valley length.

siphon   A water-filled 'U' shaped conduit with both ends submerged, and which facilitates water levels in two storage vessels, so connected, to equilibrate. This occurs naturally in some limestones, and gives rise to ebb-and-flow springs.

siphon continued   Water levels also equilibrate through an inverted 'U' shaped connector, a so-called *inverted siphon*. This is a misnomer, as a siphon mechanism *necessarily* utilizes suction (less than atmospheric pressure in the conduit).

skewness   A measure of asymmetry of a probability distribution. The coefficient of skewness, $g = \dfrac{N\sum(X_i - M)^3}{(N-1)(N-2)\sigma^3}$

for $i = 1 \ldots N$, where N is the number of data points, M is the mean, and σ is the standard deviation. Most hydrologic data sets, such as floods, droughts, rainfall amounts and intensities, *etc.* are positively skewed; the mode is less than the mean.

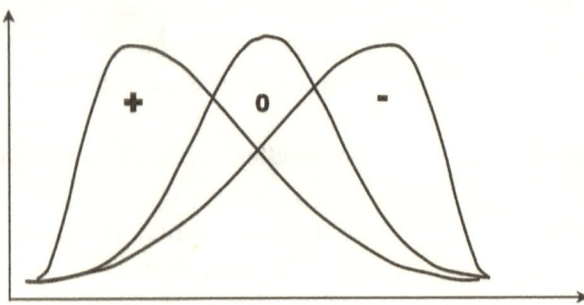

skin depth A measure of the penetration depth of electromagnetic radiation, as measured from the ground surface. Conventionally, the skin depth is the depth, in metres, at which the electromagnetic field is 37% of the surface intensity, and is determined as

$$d(m) = \sqrt{\frac{\rho}{f}}$$, where ρ is the resistivity in $\Omega.m$, and f is the frequency in Hz. *NB:* In a homogeneous halfspace the field intensity decreases exponentially with depth.

skin effect (1) The difficulty of getting enough current into the ground to measure its apparent resistivity when the ground surface is so dry that the surface electrical resistance is huge.
(2) The linear drawdown in a pumped well caused by the well itself, as opposed to the formation drawdown. The total drawdown in a pumped well = the formation head loss + the skin effect + a turbulent (non-linear) head loss. If the effective radius of a well > the borehole radius, then the skin effect is positive, but if there is encrustation

of the screen and/or poor development, then the effective radius may be small, and the skin effect negative.

skin factor   The product of the skin effect and $\dfrac{2\pi T}{Q}$

sling psychrometer   This is also known as a 'Whirling psychrometer', or 'whirling hygrometer'. A two thermometer (wet and dry bulb) 'pocket instrument' for estimating relative humidity. Tables and calculators are available for converting the temperatures to R.H. There is a slight error due to variation between the wet bulb temperature, and the degree of its air exposure.

slope-area method   Indirect measurement of peak discharge in a channel, usually employing Manning's equation, and possibly modified to take account of non-uniform flow. The method is widely used for estimating flash floods in ungauged catchments.

slug test   A down-hole single-dose infiltration test in which the recession characteristics of introduced water are used to estimate very low transmissivities, such as when there is insufficient yield for a normal aquifer test. Slug tests are influenced more by the disturbed formation in the immediate vicinity of the borehole than by the true nature of the aquifer further from the well, and hence should not be regarded as a cheap alternative to aquifer testing.

small catchment   A catchment in which the rainfall distribution may be assumed (without undue error) to be homogeneous. Arbitrarily defined as less than 25 Km².

smooth pipe   The inside surface of any pipe in which the roughness elevations do not protrude through the viscous sub-layer, and hence do not influence flow, in which case the roughness coefficient $f = \dfrac{0.316}{R_N^{0.25}}$ for $2300 < R_N$

	$< 10^5$, where R_N is the Reynolds number, and f is the friction factor.
SMOW	Standard mean ocean water, or V-SMOW (Vienna-SMOW). The universal reference composition against which $\delta^{18}O$ and δD is either enriched or depleted.
snout	The leading edge of a glacier.
snow balance	An instrument for measuring the weight of a cut cylinder of flow, and hence of estimating its water equivalent.
snow line	The altitude which separates permanent from seasonally lying snow. It does not include local shaded anomalies. The snow-line is strongly controlled by latitude, and to a lesser extent by the regional availability of moisture.
snow markers	Various designs of colour-banded stakes, sometimes used in clusters, to facilitate the measurement of snow thickness from a distance. The product of snow thickness and relative density gives the precipitation.
snowmelt	Streams close to glaciers and snowfields tend to show diurnal oscillations in flow. Otherwise, snowmelt normally results in a downstream seasonal hydrograph similar to that of ground-water (baseflow). Warm winds, of Fohn or Chinook type, can result in remarkably rapid or even catastrophic snowmelt.
snowpack	The seasonal or over-year accumulation of snow, generally in high relief areas, which is important to quantify for the purposes of avalanche and flood warning, and spring snowmelt assessment. Normally the water equivalent of a snowpack requires snow-pillow instrumentation in high/remote areas, and

hence instrumental data collection is limited. If the snow is not too thick, *and* if the natural background γ-radiation is known, *and* if there is no cloud cover, then it is possible to estimate the snowpack of a basin by measuring the attenuation of γ-radiation by its absorption the snow. This can be done by either airborne or satellite sensors.

snow pellets = Soft hail. Precipitation of soft white ice grains, as opposed to hail, which is harder, and sometimes clear.

Snow-pillow An instrument for measuring the water equivalent of snow that collects on a flat surface collection area. There are several designs, and since instrumentation is often inaccessible in heavily snowed areas, modern snow-pillows usually deliver the data by telemetry.

snow water content The amount of water *(sensuo stricto)*, in mm, held within a column of snow, but not including the snow itself. This is determined calorimetrically. Not to be confused with snow water equivalent.

snow water equivalent The amount of water, in mm, which would be obtained by melting a column of snow, and including any water that the snow may contain. In parts of America there has been no significant change in snow thickness since the 1970s, but the snow has become 'dryer', with a 14% decrease in snow water equivalent.

sodic soil A soil in which a high proportion of the exchangeable sites are occupied by Na^+. In such soils, thick layers of oriented water molecules are bound, resulting in a hard texture with strong physical properties.

sodium adsorption ratio $SAR = \dfrac{cNa^+}{\sqrt{\dfrac{1}{2}\left(cMg^{2+} + cCa^{2+}\right)}}$ where the concentrations, cX, are in meq.l^{-1}. SAR is a parameter used to assess the risk of excess sodium in irrigation water.

soffit The soffit level, that is the minimum clearance under the deck of a bridge over a river, is arguably its most important design parameter. The most comprehensive expression for the soffit level, including coastal rivers, is: $C = (M+S+D+T+TS)*FS$, where C is the total clearance to be allowed, relative to the mean base-flow elevation of the river, M is the maximum observed or calculated flood level above the mean base-flow level, S is the superelevation, D is the allowance for floating débris (often underestimated), T is the tidal half-amplitude of the proxigean spring tide, TS is the allowance for tidal surge, and FS is the assumed 'factor of safety'.

software As opposed to hardware: the human activities and intellectual products required to initiate and maintain any system, such as
(1) The programmes and applications used in computing.
(2) Training, education, motivation, and communication with personnel in any water supply or water quality project.

soft water An imprecise term for water that lathers easily with soap, where there is little or no Ca and Mg carbonate or sulphate present. *See* **water softening**.

soil detention storage Any transient amount of soil moisture in *excess* of the field capacity. The soil detention storage varies from a maximum under saturated conditions, and decreases by gravitational drainage until it reaches zero at field capacity.

soil moisture The total moisture content (pellicular and gravitational) of a soil.

soil moisture deficit The extent to which soil moisture, in mm, falls *below* field capacity. *cf.* 'soil detention storage'. The soil loses moisture when evapotranspiration exceeds infiltration. It is an approx-imate indication of seasonal transpiration stress, and of the amount of irrigation required for optimum growing conditions.

soil profile Soils are immensely varied, but in general any vertical section can be classified into various sub-horizontal zones, each of which may be subdivided on the basis of texture. From the uppermost, or 'A' horizon, soluble minerals are leached down to the 'B' horizon, where they are re-precipitated. The 'A' horizon is commonly highly pervious and rich in humus, with a granular or fibrous friable texture. The 'B' horizon is usually clay rich, relatively impervious, and hence the limiting factor in permeability of the whole profile. The A-B interface may be sharp or gradational, but the B-C interface is nearly always a transition into weathered parent rock. The 'C' horizon tends to be gritty, and may be oxidized or reduced.

soil water *See* **soil moisture**.

solar constant The mean solar energy received at the mean earth orbital radius, as measured above the Earth's atmosphere, = 1361 $W.m^{-2}$. This includes radiation at all wavelengths including infra-red and ultra-violet. It is not quite as 'constant' as the name implies. There is an apparent seasonal variation of about 7% caused by the Earth's elliptical orbit, as well as minor changes in the Sun's radiant output, of ~0.1%. *cf.* 'global radiation'.

solifluction Literally, 'soil flow'. The gravity mass flow of a soil, especially enhanced by numerous cycles of freezing and thawing. This is most common in higher latitudes, and may occur on slopes as low as 2 or 3 degrees.

solubility product For any solute at equilibrium with water, (at saturation), and which ionises, the solubility product is the product of ionic activities. For example, the reversible dissolution of gypsum may be written as: $CaSO_4.2H_2O \rightleftharpoons Ca^{2+} + SO_4^{2-} + 2H_2O$. The solubility constant is written in terms of products (numerator) and 'reactants' (denominator):

$$= \frac{[Ca^{2+}].[SO_4^{2-}].[H_2O]^2}{CaSO_4.2H_2O}.$$ But both the water and solid phase activities are taken as unity, so the solubility product reduces to: $K = [Ca^{2+}].[SO_4^{2-}]$.

solute Any phase that can be permanently dissolved in a solvent.

solute potential *See* **osmotic potential**.

solute transport equation $\frac{\partial c}{\partial t} = D\frac{\partial^2 c}{\partial x^2} + \bar{v}\frac{\partial c}{\partial x} \pm J$ The rate of change of concentration comprises, respectively, a diffusion-dispersion term, an advective term, and a reaction (J) term. The latter may be positive or negative according to whether it is a source or sink, and is usually the most difficult term to define because it may involve any combination of processes such as: adsorption, desorption, decay, dissolution, precipitation, kinetic and temperature effects, competing and biological reactions.

solution groove *See* **rillenkarren**.

solution pocket ≈ **cupola**. A karstic cavity, usually dome-shaped, but sometimes rounded wedge or irregular shaped. These false flowpaths are 'dead ends', probably dissolved

Dictionary of Hydrology and Water Resources

away by mixing corrosion, such as by seepage along a fracture intersecting a saturated channel.

solvation Chemical degradation of a plastic-based casing by some organic solvent.

sonic log = **acoustic log** = **velocity log**. A borehole logging device, seldom used in groundwater, but much used in the oil industry. It is obtained from an axially centred probe which is lowered down a borehole, and which transmits and receives ultrasonic pulses. Interpretation utilises an adaptation of seismic refraction theory. It is inferior to neutron logging in assessing porosity, but is useful in detecting static water levels, fractures and cavities. The latter causes 'ringing', and is a means of testing the integrity of cement bonds behind casing.

SOR Successive over-relaxation. In finite difference computer models, such as to determine the piezometric configuration of an aquifer, the solution is reached by a potentially lengthy process of iteration. SOR is a means of reducing the number of iterations so that the computation rapidly converges to the solution. However, if you try to be too smart by converging too fast, there is some danger of the iteration becoming unstable, in which case the method fails to reach a solution at all.

sorbtion If you are not sure whether a solute is adsorbed or absorbed, simply refer to it as 'sorbed'. Sorbtion is just a general term for the transfer of a solute from solution to solid.

sorbtion paramater = ion exchange distribution coefficient, K_d (units of $ml.g^{-1}$). It is the ratio of adsorbed species concentration to the solute concentration under specified conditions. It is empirically derived, and indicates the tendency for a solute to be adsorbed onto a solid substrate such as clay particles. K_d values may vary from 0 to about 1000, but are notoriously difficult to measure definitively because K_d is highly sensitive to the variables: grain size (surface area of substrate) and composition, mixing efficiency, residence time, pH, Eh and water chemistry.

sour condensate Aqueous effluent from petroleum works, etc. Water which is highly enriched in H_2S, NH_3, phenols and BOD.

spate A state of flood. A flooding river is said to be 'in spate'.

spatial error Map parameters based upon a small number of sampling points, such as raingauge, borehole or chemical sampling points, tend to yield absurd contours. The contouring software does not take sample error or contour feasibility into account, such that some points are taken as maxima or minima when they are nothing of the kind. What is the probability of any of a dozen raingauges being placed at exactly the highest or lowest rainfall occurrence?—Essentially nil, yet that is what the contouring package implicitly assumes. Similar ground-water arguments prevail.

spear point A rigid screen with a sharp cone at one end, capable of being driven into unconsolidated sediment. Useful for shallow de-watering purposes or for temporary piezometric observation.

specific area The mean area (in km^2) represented by a raingauge within a network. WMO recommended values for national networks vary from 10^2 to 10^4 km^2, depending upon the type of terrain. Naturally, much lower values are required for research work.

specific capacity = **specific well capacity**. A water-well performance measure, consisting of the yield per unit drawdown: $m^2.day^{-1}$. Note that these are also the units of transmissivity. S.C. and T are closely related but not the same. Usually, clients are much more interested in S.C. than in hydraulic properties of the aquifer, although the latter are much more meaningful from a water resource viewpoint. If T has not been measured directly, it can be estimated from the S.C. and well efficiency.

specific conductance *See* **electrical conductivity**.

specific discharge A characteristic of a catchment in a given climate; the mean discharge rate per unit area. It varies from <0.1 $l.s^{-1}.km^2$ in hyperarid areas to about 35 $l.s^{-1}.km^2$ in the case of the Amazon. The specific discharge of small or arid catchments is too variable to be of much significance.

specific energy The sum of the mean dynamic stream depth and the velocity head. Hence, the total energy in terms of head, relative to the stream bed.

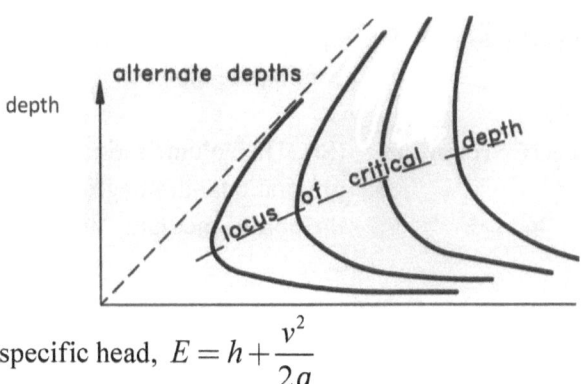

specific head, $E = h + \dfrac{v^2}{2g}$

specific gravity = **relative density**. The ratio of a volume of substance to the same volume of pure water at 4°C and at normal atmospheric pressure.

specific head = Specific energy. The sum of instantaneous and kinetic energy heads of open channel flow.

$$E = h + \frac{v^2}{2g}, \text{ or more correctly, } E = h + \alpha\frac{v^2}{2g},$$

where α varies from about 1 for smooth regular channels, to 2.0 for strongly frictional channels. For any given reach, discharge, specific head and depth of flow can be related in a family of specific energy curves, upon which the regimes of subcritical, critical and supercritical flow can be delineated.

specific heat (capacity) The energy required to raise the unit mass of a substance by 1°C. The specific heat has units of kJ.kg^{-1}.°C^{-1}; not to be confused with the *volumetric heat capacity*, which has units of kJ.m^{-3}.°C^{-1}. Typical values of specific heat capacity are: water, 4.2; ice, 2.1; air, 1.0; and soils, 1 to 3 kJ.kg^{-1}.°C^{-1}

specific humidity (q), for a given volume, q is the mass of water vapour per mass of moist air $q = \dfrac{m_w}{m_w + m_a}$ [in g/Kg] = r$_w$/r. r

specific retention (Sr). The volume ratio of water that is retained after gravitational drainage, to that of the total volume of the porous medium. Vr/Vt, where Vr + Vg = Vt.

specific speed $n_s = \dfrac{N_e \sqrt{P}}{h^{1.25}}$, where N_e is revolutions per minute, P is the power in kW, and h is the delivered head. Turbines of a particular design will have a constant specific speed, independent of size, associated with a particular range of peak efficiency. *NB:* Metric and imperial specific speeds are different.

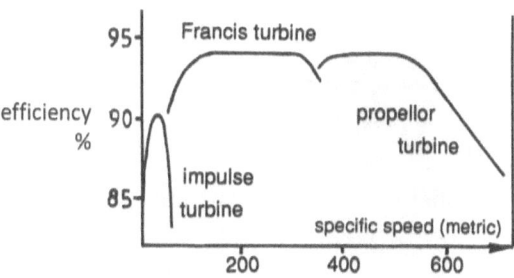

After Moody L.F., in Handbook of Applied hydraulics, (1952)

specific storativity (S_s = Specific storage. The amount of stored water released from a unit volume of aquifer per unit decline in head. In a confined aquifer S_s comprises two components: water expansion (due to decompression), and aquifer matrix compaction.

S_s *continued* The aquifer matrix compaction is caused by greater intergranular pressure resulting from decreased hydrostatic pressure. Thus $S_s = \rho_w g (n\beta_w + \beta_a)$ where ρ_w is the water density, g is gravity, n is porosity, and βs are the compressibilities of water and aquifer. The storativity, S, of an aquifer is a more widely used parameter: S = S_s.D, where D is the effective thickness of the aquifer.

specific stream power The energy available for erosion and overcoming friction, per unit area of streambed:

$$\omega = \frac{\Omega}{w} = \tau_0.\bar{v},$$ where Ω is the gross stream power

(sometimes loosely referred to as just the 'stream power'), w is the streambed area, τ_0 is the downslope shear component of water at the the bed-stream interface, per unit area (= $\rho g h.\sin \alpha$, where h is the

	depth, ρ is the fluid density, and α is the slope), and $\overline{v}$ is the mean velocity of flow.
specific surface	The surface area per unit volume of soil or sediment.
specific water capacity	The change in water content per unit of osmotic potential.
specific weight	Density in units of weight per unit volume. $\gamma = \rho.g$ in $N.m^{-3}$, or $kg.m^{-2}.s^{-2}$. Not to be confused with specific gravity, which is dimensionless (s.g. = 1 for water).
specific yield	(Sy). The volume ratio of water that drains by gravity, to that of the total volume of the porous medium. Vg/Vt, *cf.* 'specific retention'. In fine sediments it may require a very long time (years) for the cumulative gravitational drainage to approach the specific yield.
speed limit	In order to minimize bank erosion, and to protect riparian ecology, many river authorities impose a speed limit upon boats, typically of between 4 and 6 knots. In reality the dominant factor in bank erosion from the wake of boats is displacement rather than speed. *See also* **transmitted wave energy** and **displacement**.
speleology	The study of caves and their formation. A subject closely related to, but not synonymous with, **karst hydrology**.
speleogenetics	All physical and hydrochemical processes which contribute to the formation of caves and endokarstic phenomena.
speleothem	A 'bucket term' for all forms of mineral precipitation, but most commonly associated with endokarstic development.

spill Accidental pollution or, more specifically, the uncontrolled flow from a reservoir.

spillway A designed channel to convey overtopping flow from a reservoir. It may be the top of a dam, or a channel to one side of it, or a tunnel from a bell-mouthed vertical intake in the reservoir. The latter is a 'glory-hole spillway'.

spin test A subjective field test in which an impellor or anemometer type current meter is spun by blowing on the vanes. Any sluggish-ness to respond suggests a bearing friction problem, and a need to service and / or recalibrate the instrument.

spoil heap A heap of excavated soil and rock from a dug well or reservoir. Useful information about an aquifer can sometimes be gleaned from a detailed examination of unweathered spoil heaps.

SPOT The French multi-spectral satellite imaging system. The images have a pixel resolution of about 30 m x 30 m.

spouting velocity The velocity gained by loss of head (excluding friction).
ie. $\sqrt{2gh}$.

spring A groundwater exit point. All natural springs are structurally or lithologically controlled. Numerous types of springs include: subaerial, submarine, 'ebb-and-flow', vauclusian, depression, perennial, ephemeral, acidic, 'soda', mound, and alkaline.

spring tide A maximum tide caused by the coincidence of solar and lunar gravitation. The 'spring range' is the corresponding tidal amplitude. The proxigean spring tide is an unusually high tide which occurs when the moon's orbit brings it unusually close to Earth, a little less than once a year. Obviously, the new moon brings a higher tide than the full moon.

spudding in — Drillers' slang for starting to drill a well, when the bit first starts to cut the ground.

stabilization
1) The process of reducing a contaminant's mobility by binding with clay, cement, resin etc. = chemical fixation.
2) Ways of maintaining a stable embankment, including de-watering by toe drains or by pumped interception of ground-water.

stability of flotation $= h_m . \tan \theta$, where h_m is the metacentric height, and θ is the inclination of the floating body from the vertical.

stage — The river or lake level measured as height above an arbitrary datum (at or beneath the lowest level of the river or lake cross-section).

stagnation point — A point in any flownet (surface or groundwater) in which the water velocity is zero (it may rotate but does not advect).

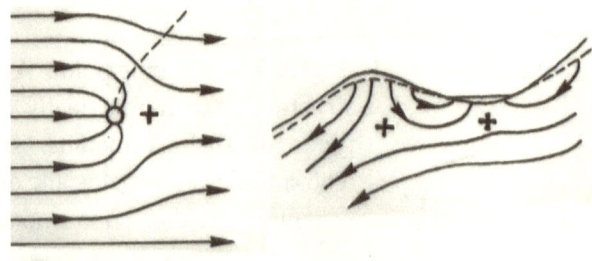

borehole extraction

standard — ... *of water quality* ... A legally enforceable (HPL), or recommended maximum (HDL) concentration in solution. Standards are pragmatic limits, set in the

Dictionary of Hydrology and Water Resources

light of 'safety margins', cost-benefit and risk benefit considerations. A 'standard' is not a 'criterion'.

standard deviation (σ) A measure of the scatter of a data set about its mean value. $\sigma = \sqrt{\dfrac{\sum (x_i - M)^2}{N}}$, where M is the mean, and N is the number of data points.

standard error The square root of the variance of a sample distribution.

standard gauge The standard U.S. weather bureau design of raingauge (8 inches in diameter).

standardized precipitation index The SPI is an index of drought severity based upon the rainfall frequency distribution, and from that, the cumulative probability distribution. Unlike the PDSI and CMI, it is valid for all time-scales. It is somewhat complex, but easier to compute than most other drought indices. (Software is readily available).

standard project flood A rather imprecise design flood, defined as the worst case under normal conditions, but excluding extremely rare and highly improbable maxima.

standing waves As the name implies; an apparently stationary wave. The classic example is a hydraulic jump, but a standing wave can occur in several other ways, including the interplay of two opposing wave trains, fast flow over an obstruction, or by the dissipation of kinetic energy in a tail race.

standpipe = **standpost**. Generally, a vertical pipe and supporting structure, with one or more water taps. Also, a narrow-bore observation well for monitoring groundwater levels. A 'public standpipe' is for general use, a 'neighbourhood standpipe' is for a more restricted user group, and a 'yard tap' is for specific family use. Standpipe is a rather vague and overused term with numerous connotations.

Stanger's Laws ... of water resources. Blindingly obvious, self-evident and basic practicalities of hydrogeology, hydrology, and water resources management, applicable everywhere but particularly in under-resourced developing countries. They are:
Law 1: You can't manage what you don't understand.
Law 2: You can't understand it unless you measure it.
Law 3: You can't measure it by sitting in an office.
These cannot be by-passed by computer modelling and data synthesis, use of empirical equations, remote sensing, *etc*.

Static Water Level (SWL). At least, a *relatively* static point groundwater level, as opposed to dynamic or recovering water levels. There is probably no such thing as a totally static piezometric surface.

stationarity An implicit assumption in most statistical analysis, particularly time-series of hydrological events, is that there is no significant drift of probability distribution either whilst the data was being collected, or within the domain of projection. That is, that all data points define a stable and valid distribution. The era of climate-change effectively scuppers that concept!

steady flow Invariant discharge; depth and velocity remaining constant at a given point. Essentially true for the duration of most flow measurements, and the underlying assumption in the application of almost all discharge equations.

steady-state Although almost nothing is steady-state in the strict sense, many situations are analytically tractable by the assumption of steady-state conditions, for instance in catchment water balances, streamflow or groundwater flow. In particular, aquifer tests are analysed either by transient or steady-state equations. If a confined aquifer is pumped for many hours or several days, then the cone of depression may reach quasi steady-state conditions, in which case a simple equation for drawdown may suffice.

stemflow Rainfall that dribbles down a canopy and tree trunk to the ground. It is seldom more than 1% of the total rainfall, even in a rainforest, but it is sometimes of significance in transporting aerosol pollutants and / or biomass nutrients from the tree to the soil.

stem psychrometer A paired thermocouple instrument for continuously monitoring the (negative) vascular pressure, or water potential, in plants. The base or 'rest' reading is before dawn. During the day the psychrometer records the plants' diurnal stress cycle, integrating the aggregate effects of soil moisture, solar radiation, stomatal sensitivity to light of varying spectra, humidity, wind speed, temperature and plant physiology.

stepped tariff Few places in the world can afford the luxury of cheap water for everyone. Universally cheap water invariably results in some degree of wasted water, yet social concern dictates that the poorest of society should get a measure of cheap water. The compromise is to price public water supply according to a stepped tariff, where the pegged level of each step is logically determined from a storage-cost curve. The number of steps varies from two to 'continuously variable'. In heterogeneous- income societies it is common to have the first step free, as in the example below, from Johannesburg. Where water supply is limited, high

demand can be discouraged by the highest usage being linked to the steepest tariff increments.
example of a stepped tariff, from Johannesburg

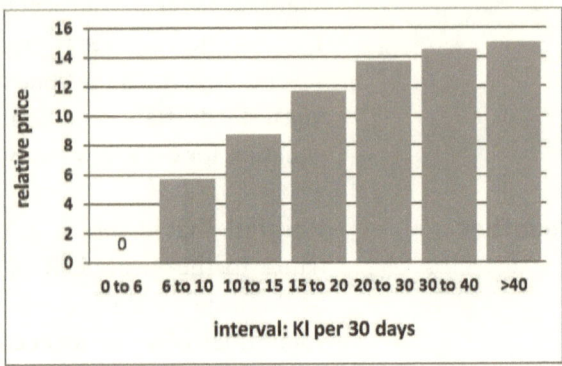

stilling basin	(1) A reach immediately *downstream* of a dam or outflow, designed to dissipate the water's kinetic energy.
	(2) A bank inlet or other protected area of river flow immediately *upstream* of a weir, with low or zero water velocity, constructed to permit accurate measurement of stage, without loss of velocity head.
	(3) A large diameter bore with a damped inlet in hydraulic continuity with a tidal river, estuary, harbour, *etc.*, designed to facilitate measurement of long-period water level fluctuations with minimal interference from wind-waves, wakes, and other short-term 'noise'.
stochastic	A stochastic model or simulation is one based upon statistical probability, and therefore incorporates some element of random chance, as in the estimation of flood peaks. The adjective 'stochastic' in often used in opposition to the alternative of 'deterministic', in which all the variables are measured. Stochastic models mainly occur in relation to hydrology, but economists have also hi-jacked the term.

Dictionary of Hydrology and Water Resources

storm surge　　Most extreme floods are caused by a combination of offshore and onshore storm conditions in which a storm surge is compounded by other factors. The maximum water level in coastal and near-coastal reaches is the high tide + the wind-driven surge + the pressure surge + the riverine flood wave (moving in the opposing direction). The storm surge is the sum of the offshore surge, caused by low atmospheric pressure lifting sea level, and the wind-driven surge. The wind surge may be up to 10 times larger than the pressure surge. Obviously, the potential damage is exacerbated by land subsidence and climate-change related factors such as sea-level rise and intensification of storm rainfall—a reality which many planners do not yet seem to have grasped.

storm track　　The direction of drift of a storm cell. This can influence the shape of the resulting hydrograph. A storm track moving upstream attenuates the hydrograph, whilst a storm track moving downstream tends to sharpen the flood peak.

Strange's Tables　　In 1928 W.L. Strange published tables for estimating the monsoon runoff coefficient in a few small catchments in the Western Ghats, near Mumbai, India. Despite general usage of these tables being thoroughly discredited decades ago, Strange's tables continue to be misapplied in lower rainfall areas all over India, resulting in billions of rupees totally wasted by lazy hydraulic engineers in 'designing' useless dams that will never fill with water. It is indeed strange that anyone still believes this balderdash. If you see a copy, burn it. *See also* **Stanger's Laws**.

stabilization　　Numerous methods are available to stabilize bank scour. These include stabilization with geotextiles, steel piling, willow planting, rip rap, *etc.* Ultimately it is in the nature of rivers to be unstable.

step drawdown test A type of aquifer test in which measures drawdown *vs.* time in the pumped well for incrementally varied pumping rates. This drawdown comprises a linear formation loss, associated with Darcian flow, and a well loss which is mostly, but not entirely, non-linear.

step test continued The formation loss is expressed as $s = BQ + CQ^n$, where s is the drawdown, Q is the discharge rate, B and C are loss coefficients, and n is normally taken as 2, although it can vary from 1.5 to 3.5. The transmissivity can be crudely estimated from a step test.

Stern layer The innermost layer of the 'electrical double layer' of aquatic ions, which is found immediately adjacent to a charged particulate surface. The Stern layer contains uniformly charged ions (usually cations), and is very strongly adsorbed to the surface.

Stiff diagram A graph comparing the relative distribution of aqueous cations and anions in terms of milli-equivalents, for example:

```
            + meq.l-1     0      meq.l-1 -
    Na+ + K+ ─────────┐         ┌───── Cl-
        Ca2+ ─────────┤         ├───── CO3 2- + HCO3-
        Mg2+ ─────────┘         └───── SO4 2-
              example of a Stiff diagram
```

stilling basin (1) ... *in a well* ... = a gauge well. An air-vented standpipe or chamber in or hydraulically connected to a stream in such a way as to facilitate stage measurement. It has short-term stage fluctuations damped out, and with effectively zero kinetic head loss $\left(\dfrac{v^2}{2g}\right)$.

(2) ... *behind a weir or spillway*... A pool, usually concrete lined, and sometimes containing chute blocks, baffle piers, dragons' teeth, or other devices to dissipate kinetic energy.

stochastic Pertaining to a statistical distribution of events rather than the magnitude of any particular event. For example, the stochastic approach to flooding is to consider a catchment as a 'black box' generator of floods. If enough floods are studied, their probability density function may be computed, and hence their frequency, although it is not possible to predict the magnitude of the next flood. *cf.* **deterministic**.

stock water Livestock tend to have a much higher tolerance to salt than humans. Some typical examples of maximum TDS (mg.l⁻¹) for stock water are:

poultry	2900 to 3500
pigs	4300 to 4500
dairy cattle	6000 to 7100
beef cattle	7000 to 10,000
sheep on dry feed	13,000 to 14,000
Ewes with lambs	4500 to 10,000

stoichiometry A chemical reaction occurring in an exact weight proportions. In a stoichiometrically balanced reaction there are no 'unreacted leftovers'.

Stoke's Law At constant (terminal) velocity v, a sphere of radius r, moving through a fluid with a coefficient of viscosity η, under laminar flow conditions, experiences a viscous force (drag) $F = 6\pi r \eta \varpi$. From this it can be deduced that $v = \dfrac{2\Delta\rho g r^2}{9\eta}$ where $\Delta\rho$ is the differential specific gravity (solid-liquid). Although defined idealistically, the viscous forces described by Stoke's Law have important consequences in hydrodynamics, groundwater flow, and elsewhere.

stoma — Minute narrow-mouthed cavities in leaf and sometimes stem surface which facilitate and control gas exchange in general, and transpiration in particular. (*plural:* **stomata**).

stone forest — An extraordinary geomorphic phenomenon peculiar to certain classic karst regions of Southern China. Stone forests consist of limestone valley floors subjected to intense karstic dissolution along numerous joints. The result is innumerable stone pillars separated by rock debris and silt. The resulting hydraulic conductivity is much greater in the vertical than in the horizontal. *See* **fenglin** and **fengcong**.

storage coefficient — = Storativity, 'S'. The volume of groundwater released or taken into storage per unit plan area of aquifer per unit change of head. It is dimensionless. In the case of unconfined conditions, the storage coefficient is the specific yield x aquifer thickness.

storage gauge — A raingauge with a large collecting sump, used in areas of difficult access and infrequent monitoring.

storativity ratio — 'ω'. The fraction of total storage which is secondary porosity. It is a measure of the relative importance of fracture to matrix storage. $\omega = \dfrac{S_f}{S_m + S_f}$, where S is the storage, and the subscripts refer to the matrix and fractures.

storm efficiency The ratio of actual precipitation to the total moisture content of a column of air.

storm lane The ground locus of preferred storm tracks. This is particularly noticeable in some mountainous areas with strongly prevailing winds.

storm seepage Infiltrated storm water that moves laterally through the unsaturated zone.

storm water ≈ **quickflow**. Urban stormwater was once thought of as a problem to be disposed of as quickly as possible. It is now regarded as a significant water resource.

stratified aquifer An aquifer in which vertical variations in the hydraulic conductivity cause alternate high and low yielding horizons. Most aquifers are stratified to some extent, hence a range of hydraulic conductivities may contribute to the overall transmissivity.

stratified flow A flow regime in which less dense water overlies denser water. Fresh or warm water may overlie cold, saline or sediment- laden water.

stream degradation Erosion of a stream bed. There are several possible causes, including river capture, tectonic rejuvenation, and upstream reduction of the sediment load (such as by constructing a dam).

stream function $\Psi_{x,y}$. Notation for the flowlines in a flownet, defined in terms of x and y velocities:

$$v_x = \frac{\partial \Psi}{\partial y}, \text{ and } v_y = \frac{\partial \Psi}{\partial x}$$

stream head hollow The head of a valley where surface runoff converges.

streamline A hypothetical line of flow of a 'point' of water. Streamlines are at all times perpendicular to isopotentials, and together with isopotentials, can be used to construct a flow net. In a long profile of a stream, both the surface and any impervious stream beds are the bounding streamlines.

stream order A numerical hierarchy for classifying the component reaches of a drainage system. Normally, the order is 1 for the head-waters, and increases methodically below key confluences, such that the highest order occurs at the mouth of the river. One exception is Gravelius's system in which the main channel is everywhere of order 1, whilst successively minor tributaries are of higher order. There are at least 5 different methods of ordering, so it is essential to quote both the stream order and the system being used.

stream power Ω, the rate of potential energy loss per unit length of channel. This is an important geomorphological parameter. In minimizing the energy loss, Ω, influences the stream slope, cross section and planform. $\Omega = \rho g Q S$ in W.m^{-1}, where ρ is the density, g is gravity, Q is the discharge rate, and S is the energy gradient ($\approx$ hydraulic gradient).

stream-tube The area bounded by two specified streamlines of a coherent flownet.

stress Force per unit area. *See also* **transpiration stress**.

Strickler's equation An empirical equation for estimating Manning's roughness coefficient, n: n= 0.0474 $D_{50}^{1/6}$, where D_{50} is the median particle diameter, in *metres*, of the streambed sediments (or of the armoured layer if armouring occurs). The use of 'D_{50}' has been challenged, with some authors preferring 'D_{mean}' or '$1.25D_{50}$'.

subartesian Confined non-overflowing groundwater conditions.

subcritical flow Streamflow in which the Froude number, Fr < 1. In subcritical flow a wave can travel upstream against the flow, or the hydraulic effects of a downstream change of stage can be detected upstream. There is no unique stage-discharge relationship upstream of a subcritical reach. *cf.* **supercritical flow**.

subglacial Referring to a range of fluvial and ice-flow processes that occur beneath a glacier; as opposed to englacial, supraglacial or proglacial: within, above or in front of a glacier.

subhumid A climatic condition of sufficient precipitation to maintain perennial grass, but not forest, such as prairies, or steppe.

subirrigation Irrigation by pipe from below the ground surface, or by maintaining a shallow water table.

sublimation The phase change from solid to vapour, or vice versa, especially, relating to evaporation directly from ice. In the upper atmosphere, sublimation from vapour to ice gives rise to wispy cirrus clouds.

submerged culvert A culvert flowing full-bore. *NB:* Submergence of the inlet does not guarantee fully submerged conditions.

subsequent stream A stream or river flowing 'down dip' in accordance with the local geological structure.

subsidence Depression of the land surface may be sudden, as in the case of a karstic or mining collapse feature. However, most of the world's serious subsidence problems take place over decades, and are caused by the dewatering of compressible aquifers.

subsurface runoff Rapid interflow.

subvariable flow Of a spring: The condition in which $0.25\bar{q} < (q_{max} - q_{min}) < \bar{q}$, where $\bar{q}$ is the mean discharge.

suction *See* **matric potential**.

Sudd (Arabic) Any major obstruction to flow, but especially the enormous swamp of the Upper White Nile.

suffosion If a doline develops beneath a sedimentary cover, then fine sediments may be dissolved or washed into the karst formation, resulting in a volume decrease in the overlying sediment. Suffosion is the subsidence that results from this process. Suffosion dolines are a common feature in glacial till where it overlies karstic limestone.

SUH The Synthetic Unit Hydrograph. When the unit hydrograph shape for a catchment is unknown, an approximation, the SUH, can be synthesized from the stream geomorphology. It isn't very accurate, but is arguably better than nothing.

suitcase sample Drillers' term for drill cuttings which indicate the top of some major impervious or low-yielding formation. The implication is that it is pointless to continue drilling for water, and one may as well pack up and go drill elsewhere. The most obvious suitcase samples are the first cuttings of crystalline basement.

sullage Sediment deposited by a stream (which is most obvious after overbank flow).

sulphur spring A spring with a noticeable smell of H_2S. This is characteristic of reducing conditions, and does not necessarily correspond to a high concentration of sulphur as sulphate.

sunken pan An evaporation pan flush with the ground level to minimise aerodynamic anomalies. Care needs to be taken to minimise other measurement errors due to dust or sand displacement of water, and thermal conduction against the ground.

supercooled water Water which remains liquid despite being below its equilibrium freezing point. Cloud and high-altitude rain droplets are often supercooled to around -10 to -30°C.

supercritical flow A streamflow condition for which Fr > 1, in which case a wave cannot propagate upstream against the fast-flowing current. The stage-discharge relationship is unique immediately up-stream of a supercritical reach. Hence, if available, such a site is favoured for streamflow measurement. *cf.* **subcritical flow**.

superelevation The differential water level in a river cross-section due to centrifugal force. $Z = \dfrac{v^2 b}{r_c \cdot g}$ where v is the mean velocity, r_c is the mean radius of curvature, b is the stream width, and g is the gravitational constant.

superheated water Geothermal water that is heated to higher than normal boiling point due to rapid upwelling from depth, and which is therefore not at thermodynamic equilibrium with the surface pressure. In theory the temperature of superheated water can be anywhere between 100°C and 374°C (the critical temperature). Some hydrothermal springs in Yellowstone Park are superheated to around 120±15°C, and may be recognized both by their anomalous temperatures, and by their effervescence when stirred. In hydrothermal areas be very wary of springs that could be superheated. They don't have rolling bubbles, and hence don't *appear* to be very hot - but fall in and you're dead!

superimposed drainage A drainage system which incises through superficial geology into underlying rocks of different structure. In this way, a river system which was originally concordant with the structure becomes discordant later in its evolution. This is the only explanation for some strange and improbable drainage patterns.

supersaturation The condition of a higher solute concentration than would be expected under equilibrium thermodynamic conditions. This can occur by cooling a saturated solution (or warming in the case of some sulphates and other unusual salts), by sluggish precipitation kinetics, or by a dearth of particles to facilitate crystallisation.

super-typhoon An informal term for a hurricane of upper class 4 or class 5 on the **Saffir-Simpson scale** (q.v.). As of 2013 the most intense super-typhoon reached peak wind speeds of 305 km.hr^1. As sea surface temperatures increase in response to global warming, super-typhoons are likely to break this record.

supply curve On graph of unit cost *vs.* commodity production (such as of bottled water), the supply curve is that part of the graph in excess of the minimum actual cost, which follows the marginal cost curve.

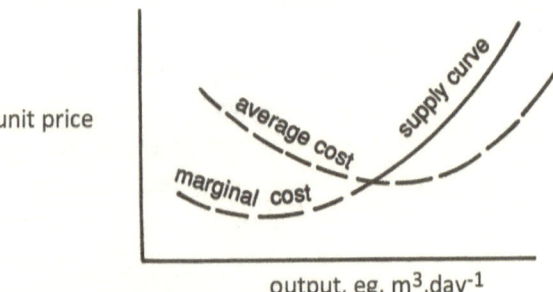

suppression	(1) *... of weirs* ... A weir is said to be suppressed if it is 'full width', since the approach velocity is assumed to be relatively uniform, as opposed to a non-suppressed weir in which the spillway is of partial stream width. The latter configuration results in complex flow around the sides of the weir, and hence head loss perpendicular to the stream axis. This is not allowed for in derivation of the flow equation, and hence requires an empirical correction factor. Suppression can also be achieved by baffles, as shown:

Plan views of suppressed and unsuppressed conditions

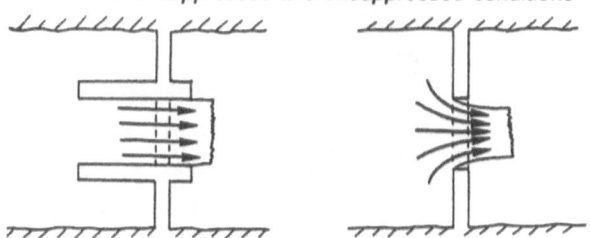

	(2) *... geophysics* ... A common feature of geophysical investigation in which the instrument response gives the impression of a simple environment, when the reality is much more complex. For example, the resolution might yield an apparent two-layer hydrogeology where the truth consists of many more layers, perhaps with gradational and irregular boundaries.
Surcharge	(1) Ephemeral dam storage over and above gross storage capacity. (2) A wide range of excuses for utility revenue raising.
surcharging	The failure of a control structure (usually a dam) to contain, control or attenuate a flood of greater than the design maximum. *NB* Failure to control does not necessarily mean structural failure.

surface casing   Wide-bore casing drilled and cemented into the near-surface, and through which the main borehole is drilled. The purpose of surface casing is to prevent collapse of uncemented surface drift, to protect the aquifer from surface inflow / pollution, and to enable pressure to be maintained in a confined aquifer.

surface detention   = **detention storage**. The film of water, including overland flow, which is left on the land surface just after rain, but not including ponded water.

surface energy   *See* **surface tension.**

surface hydrology   The study of water-related processes (mainly physical) at or near the ground surface, as opposed to hydrogeology. However there is no clear distinction between surface hydrology and various allied sub-disciplines, such as soil physics, micrometeorology and hydrography.

surface storage   Depression storage + surface detention.

surface tension   'γ', = surface energy, = free surface energy, σ. The property of a liquid surface to behave like a stretched elastic film, due to the differential molecular attraction between water and air (or any other fluids). The force of surface tension exerted along any liquid-air-solid junction of length L, is $F = 2L\gamma$. Surface tension decreases slightly with increasing temperature:

T °C	γ (in $N.m^{-1} \times 10^{-3}$)
0	75.6
5	74.9
10	74.4
20	72.7
40	69.5

surfactant A foaming agent used to disaggregate clays, and reduce surface tension. Surfactants are often used to clean up wells during development, whilst surfactant flushing is increasingly used to disperse low solubility pollutants such as alkanes and higher aromatics.

surge (1) A transient surface wave in a stream, or pressure wave in a closed conduit.
(2) A period of fast glacier movement, which may be associated with subglacial sliding. A few glaciers surge all the time, but for most, it is an ephemeral event that may occur up to 100 times faster than normal.

surge tank/pipe/chamber A large-bore vertical shaft linked to reservoir outlets, penstocks, etc. to absorb or damp out pressure surges. Without surge pipes, waterhammer could do immense damage to conduits, valves, *etc*.

surging Induced reciprocation of the water level in a well to dislodge clay and silt during well development.

suspended load ... *of a river* ... This consists of the fine-grained material that is truly suspended and transported in the water by upward components of turbulent flow. It does not include the bed load. To some extent there is continuous exchange between suspended load and bed sediments. If the sediment is 'supply limited', then suspended load < capacity load.

sustainability A concept initially made popular by an Indian guru in the 1960s, and now almost universally accepted. Since about 1990, sustainability in the use of water resources has virtually attained the status of political doctrine. Most development projects are now assessed upon the bases of at least economic viability, sustainability and environmental impact.

'Sustainability' has been described as 'a dialogue of values that defies consensual definition', yet it is the foundational principle of the triple bottom line, namely: 'indefinite economic, social and environmental activity'. As far as the water sector is concerned, ignorance, endless population growth, the 'Holy Grail of economic growth', greed, short-term expediency, competition for the last remaining water, and political inter-ference, all invariably supersede serious attempts to attain sustainability. Short-term aquifer tests may not be sufficient to determine an aquifer's sustainable yield. Most of the world's main aquifers, in all countries, are grossly oversubscribed, leaving sustainability as an unattainable pipe-dream. Many donors seem content to substitute *'sustainability for the duration of the projec*t', or *'sustainability of the management'* rather than of the resource, (i.e. *un*sustainability). 'Resilience' is overtaking 'sustainability' as the trendy buzz- word.

sustained overdraft Long-term extraction of groundwater in excess of the recharge rate. That is, 'groundwater mining'.

swale A wide shallow depression, either dug or an inter-dune depression.

swallow hole = **swallet**, = **stream sink** = **sink hole**. A common geomorphic feature of karstic terrains, in which a surface stream entirely flows into the ground (usually reappearing elsewhere as a low-elevation spring).

swamp Any water-saturated and vegetated land, usually with an abundance of standing water. *See* **marsh**, ≠ bog.

swedge block A slightly tapered down-hole ram for flaring a lead packer between telescoped sections of a bore.

symplast Part of a plant's water system, incorporating everything within a cell's membrane; the protoplast. (As opposed to apoplast).

syphon *see* **siphon**.

systematic error Systematic errors reduce accuracy without necessarily impair-ing precision and reproducibility. In theory, a systematic error is capable of being corrected by, for example, a double mass curve analysis. Examples include a false zero, collimation errors, and raingauge undercatch.

tail bay A stilling basin or pool downstream of a lock, turbine outlet or reservoir release structure.

tailings Milled or mined rock detritus from which most of the ore has been extracted. Weathering and leaching of tailings causes some of the worst examples of water pollution. Fortunately, many tailings are in remote areas where the risk to humans from elevated acidity and heavy metal concentrations is minimal.

tail race The outflow channel from a turbine or other water release point.

tailwater height The water level elevation immediately downstream of a dam or weir.

take A river 'takes' when it freezes. (Canadian term).

talus A pile of mechanically eroded coarse sediment which accumulates at the foot of a cliff or steep slope. Talus often soaks up runoff or snowmelt from high ground, and releases it lower down as a talus spring.

tannin =Tannic acid, *approximately*: $C_{76}H_{52}O_{46}$. A group of very complex organics found in natural waters and vegetation, with a capacity for precipitating proteins. Their complexation with Fe creates a dark opaque pigment which may render the water aesthetically repugnant even though it may be safe to drink.

tapping well A bore or well which just punctures the upper aquitard of a confined aquifer, but which does not penetrate the aquifer. Such wells are inefficient for water extraction, but are useful for hydrocompaction studies.

tarn = **lochan** = **llyn**. A small mountain lake, usually of glacial origin.

technobabble Obfuscation by using opaque jargon instead of plain English. GIS and IT nerds, geologists, mathematicians and statisticians are particularly prone to this, but arguably the worst offenders are members of the legal profession. *See* **legal insult**.

TDR Time domain reflectometry. Soil moisture measurement using the dielectric constant, K, of the soil as a proxy for moisture content. There are practical difficulties in measuring K directly, so it is measured indirectly by applying vhf (~0.1 GHz) electromagnetic wave theory. On larger scales, the TDR is probably the most accurate technique for measuring moisture.

TDS meter There is no such thing on Planet Earth! Many companies pretend to offer TDS meters, but they don't measure TDS. They measure the proxy parameter of electrical conductance, displayed upon an *assumed*

Dictionary of Hydrology and Water Resources

correlation between E.C. and TDS. This is a fair approximation for water of 'normal' hydrochemistry, but may be wildly inaccurate if the hydro-chemistry is in any way abnormal, as may be the cases with high salinity, or elevated acidity or alkalinity, or with abnormal anionic mix ratios. Therefore use 'TDS meter' readings with caution!

telemetry The technology of data transmission from remote sites, through either the mobile phone network or by satellite relay. The hi-tech hardware, such as solar panels and transmission mast, satellite rental and adequate instrumental maintenance, are all expensive. In most developing countries where telemetry has been pioneered, it has failed. Telemetric hydro-meteorology systems are brilliant for remote areas - provided: (a) there is a commitment to permanent funding support, (b) there is adequate in-country technical expertise to cope when things go wrong, (c) the network is well designed and deployed, (d) there is provision of maintenance, spares, and backup, and (e) there is sufficient social discipline to avoid vandalism. If any of these are lacking, forget telemetry except, maybe, for tracking turtles at sea.

telescoping The technique of drilling sections of a borehole with successively smaller bits, thus allowing successively narrower casing at greater depths. This is useful in collapsing strata, or in multiple aquifers. For example in fracking from deeper aquifers the practice of telescoping is part of the procedure used to separate freshwater aquifers from saline- or hydrocarbon-rich aquifers.

temperate ice . . . *as opposed to 'cold ice'*, generally referring to glaciers. Ice which is at the pressure melting point.

temperature There is no consistent relationship between air and surface water temperatures. However, groundwater temperatures are more conservative: In cold and temperate latitudes, the groundwater temperature ≈ mean annual air temperature + 2 ± 0.5 °C. In tropical latitudes, the groundwater temperature ≈ mean annual air temperature + 5.5 ± 2.5 °C. Apart from geothermally heated waters, shallow groundwaters have a normal temperature range from 0 to about 34 °C. The absolute temperature range of natural waters varies from -42°C in Antarctic CaCl salt lakes, to several hundred degrees in deep pressurized geothermal systems.

temperature index A rough estimate of daily snowmelt based purely upon air temperatures: $M = M_r(T_a - T_b)$ where M is the daily snowmelt, M_r is the melt rate index in mm.°C^{-1}.day^{-1}, T_a is the daily air temperature and T_b is the base temperature at which no melting occurs. $(T_a - T_b)$ is the degree of heat energy. T_a may be mean daily (about -3 to 0 °C) or maximum daily (typically -2 to +5°C).

temperature inversion *See* **inversion**.

temporary hardness = **carbonate hardness**. An archaic and imprecise measure of the carbonate salts that can be precipitated by boiling. *See* **hardness**.

tensiometer An instrument for measuring the water potential of a soil (the potential being indirectly related to the soil-water content). There are several types including the 'porous cup', 'ceramic block', or gas pressure devices.

terminal velocity A particle falling through a still fluid accelerates to its terminal velocity, after which the force of acceleration = the frictional resistance against the

Dictionary of Hydrology and Water Resources

falling particle. This is the basis for measuring the viscosity of a fluid by Stoke's Law.

terrace (1) A series of soil-retaining walls which follow the contours of a steep hillside, thus giving a stepped or 'bench' appearance. The purposes are to minimize water and soil loss, to delay drainage, and to facilitate ease of cultivation. Rice paddies in hill country are the classic example.

terrace nomenclature

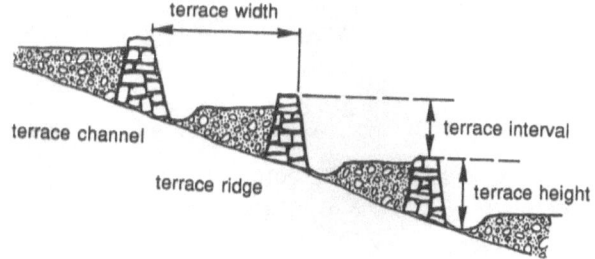

(2) A natural 'shoulder' on the side of a valley (often paired, so on both sides of a valley) caused by fluvial erosion into older floodplain sediments. In some cases, successive erosive episodes have cut a series of terraces into the same valley, in which case the highest terrace is the oldest, whilst the lowest terrace is the youngest (*NB:* the reverse of sedimentary chronology).

terrace interval The mean vertical distance between one terrace and the next, as opposed to the terrace height, which is the distance from the top of a terrace ridge to the base of the next lowest terrace channel; *see* diagram above.

terra rossa An oxidized iron-rich (and sometimes aluminium-rich) soil derived by the dissolution of limestone, and which accumulates as an insoluble residue on the surface of some flat-lying limestones. Some of the world's best wine grapes are grown on terra rossa.

terrestrial water storage In gross regional terms changes in the inter-annual water storage, resulting from floods and drought, can be assessed by satellite gravimetry. For example, the Brazilian dry-season (El Niño, 2002-3), compared with a wet season (La Niña, 2009), yielded a differential of ~625 gigatonnes of combined surface and groundwater.

thalweg (German) The long profile of a valley, which almost perfectly describes the variation in surface hydraulic gradient.

Theis equation The equation for non-equilibrium radial flow to a pumped well:

$$S = \frac{Q}{4\pi T} \int_u^\infty \frac{e^{-u}}{u} du, \text{ where } u = \frac{r^2 S}{4Tt},$$

s is the drawdown, Q is the constant discharge rate, T is the transmissivity, S is the storativity, t is the time since the start of pumping, and r is the radius from the pumping well to the observation well.

thermal conductivity The thermal conductivity of water is 0.6 W.m^{-1}.°K^{-1}, which is several times less than that of most solids, including ice. Nevertheless groundwater is an effective 'heat scavenger' in most aquifer systems.

thermal diffusivity D_t. The ratio of saturated bulk-rock thermal conductivity (K_e) to heat capacity: $\frac{K_e}{\rho.c}$, where ρ is the density, and c is the saturated bulk-rock specific hest. D_t is directly analogous to hydraulic diffusivity: $\frac{K}{S_s}$. The thermal diffusivity of aquifers typically varies from about 0.2 to 1.0 x 10^{-6} $m^2.s^{-1}$. It is a necessary parameter in the equation for heat flow in an aquifer:

$$\nabla^2 T = \frac{1}{D_t} \frac{\partial T}{\partial t}.$$

thermal neutrons   Neutrons at energies of < 0.025eV, as generated, for example, by the multiple back-scattering of fast neutrons in bore walls. Thermal neutron flux is the detected signal in neutron borehole logging, and is indicative of the total water content (+ hydrocarbons, if any) of the rock.

thermal spring   = **hot spring**. Several definitions include: ≥ 37°C (body temperature), ≥ 40°C, or ≥ (mean annual air temperature + 5°C). Causes include volcanism, deep groundwater circulation, seismicity, exothermal water-rock reactions and thermal refraction.

thermocline   A layer of any water body, which separates the upper low-density water (epilimnion) from the higher-density layer (the hypolimnion) beneath. The thermocline is defined by its relatively steep temperature gradient.

thermokarst   Depressions and erosion features caused by freezing and thawing of ice. It is somewhat of a misnomer as there is no chemical dissolution involved.

thermo-osmosis   Temperature causes slight changes in surface tension and interstitial vapour pressure which, in an under-saturated soil, can create small-scale water movement, or 'thermo-osmosis'. This is unimportant in terms of general groundwater flow, but may create localized engineering problems, such as from freezing moisture build-ups beneath an impervious surface.

thief zone   A large fracture or cavity into which drilling fluid leaks during rotary drilling, thereby causing 'loss of circulation'. Heavily fractured and/or eukarstic rocks are particularly troublesome in this respect.

Thiessen polygon   An objective means of dividing a catchment area into sub-areas, each represented by a single raingauge. A variant is the height-weighted Theissen polygon. Assessment of areal rainfall by Theissen polygons is appropriate for flat(tish) homogeneous basins, but definitely not for mountainous areas.

Thiobacillus   A genus of sulphide-oxidising chemo-lithotrophic bacteria, common in many aquatic environments, and notable in some hot springs and drainage from sulphide-rich mines. Some species have an amazing ability to thrive under 'hostile conditions' such as a pH of as little as 1, and at near boiling temperatures. Some *Thiobacillus* sp can also reduce nitrate.

thixotropy   A reversible property by which a fluid sets or becomes gelatinous when left to stand for a short time, but becomes much less viscous (more 'fluid') if stirred or agitated. This is an important property of drilling muds. Some natural clay suspensions also behave in this way, which can be very dangerous, as in the case of fluidized mudflows mobilized by seismicity.

throat   The laterally constricted part of a venturi or flume.

throughfall   (1) The amount of rainfall that falls through a vegetated area to reach the ground. It may vary from virtually zero for a light shower in dense woodland, to more than 80% for intense tropical storms in some types of rainforest. *NB*: total rainfall = interception + throughfall + stemflow.

(2) Shallow groundwater flow through a soil sub-parallel to a hillside. If the underlying rock is reasonably permeable, then the infiltrated water percolates vertically, and there is no through-flow.

tidal efficiency For an aquifer adjacent to, and in hydraulic continuity with, an open water tidal signal, the tidal efficiency is the ratio of the piezometric amplitude to that of the open water tidal amplitude. It can also be expressed in terms of the aquifer compressibility, β_a, and specific storage, S_s: $\eta_t = \dfrac{\rho_w g \beta_a}{S_s}$

tidal wave The normal twice-daily tidal oscillation of seawater, and its propagation into estuaries, rivers and aquifers; *see* **bore**. Tidal waves propagate through coastal aquifers with widely varying degrees of attenuation and lag.

Many people incorrectly refer to tsunamis (seismic sea waves) as 'tidal waves', although they are not tidal in any way.

tile drain A line sink consisting of leaky tubes, set at a shallow angle below a water table, which enhances natural gravitational drainage.

till = **boulder clay**. Glacial sediments of very heterogeneous particle size. Till is very widespread over northern regions of Europe, Asia and America. It is generally a poor aquifer with both low porosity and hydraulic conductivity.

tiltmeter An instrument for measuring the tilt, usually in micro-radians, relative to a long base line, of tens of meters, caused by loading and unloading. This finds application in assessing the long-term stability and safety of large dams.

time constant   The integration time required to take a single reading, such as a γ or neutron flux rate, or the count rate of a flow meter. Accuracy improves as the time constant increases.

time of concentration (t_c). The time taken for surface runoff to flow from the furthest headwaters to some downstream reference point or catchment outlet. Numerous empirical formulae exist for estimating 't_c'. Treat all such equations with caution. In particular, most equations for t_c are approximations for a planar basin under steady flow conditions. In reality most basins are divergent or convergent over a large part of the basin area, and display some measure of unsteady flow. In such cases the t_c may be approximated as:

$$t_c = \left(\frac{1+\alpha}{2}\right)^{\frac{1}{\beta}} \cdot \int_0^1 \frac{dx}{\left[(1-\alpha^2)x^\beta + a^2\right]^{\frac{1}{2\beta}}} \text{ where } \alpha \text{ and } \beta$$

are basin shape and friction parameters. It is doubtful whether such equations are any advantage over direct observation.

time of rise   ... *in streamflow* ... The interval between the onset of quickflow, and peakflow.

tipping bucket   An incremental flow measuring device, fitted to pluviometers. The raingauge/ pluviometer delivers the fall to a flip-flop 'double bucket' with an electrical pulse recorded at each flip. It is designed to record an increment of rainfall, such as 0.1, 0.2 or 0.5 mm. Good maintenance is required for optimal results.

tjaele   (Russian) permafrost.

TOC   Total Organic Carbon, as opposed to CO_2 species. TOC = the dissolved and particulate organic carbon, such as contribute to humic, fulvic or tannic acids. *See also* **DOC**.

toe ... *of a dam* ... The downstream basal edge of the dam.

toe drain A stable porous structure at the toe of an earth dam. Its purpose is to drain seepage water efficiently and safely without piping or significant hydraulic gradients.

'tolerable risk' A fuzzy and dubious concept of pathogen detection, most commonly required because many poor countries have neither the skills nor resources to adequately monitor potable water.

TON Total Oxidized Nitrogen. The combined concentrations of nitrate and nitrite, as opposed to ammonia and related nitrogenous ionic species.

top of casing Commonly the datum from which borehole water levels are measured. This data has to be further reduced to a common datum in order to plot the piezometric surface.

ToR 'Terms of Reference', the guidelines for project proposals, which typically specify the project rationale and background, goals and objectives, milestones and deliverables, project duration, terms and conditions for consultants. The ToR may also include other pertinent information such as the required methodology, project staffing schedule, previous studies, related projects and donors, legal constraints, logframe, *etc*.

torpedo sinker A hydrodynamic weight, designed for minimal frictional resistance, used to sink a velocity or depth probe in fast flowing water.

tortuosity The ratio of the actual flowpath in a porous medium (around particle grains), to the direct distance between two points in the granular medium.

total dissolved solids (TDS) = Dissolved solids = residue on evaporation. Despite measurement errors due to incomplete loss of water of crystallization, and to loss of volatile solids, TDS remains one of the best measures of overall water salinity. It also facilitates a check upon the accuracy of a water analysis. However, its disadvantage is that it takes several hours to determine.

total dynamic head The total hydraulic potential (taking account of velocity head) between the inlet and outlet of any rotodynamic machine.

total interception loss Stormwater re-evaporated both from the forest canopy and from litter storage without prior infiltration into the ground.

totaliser = **storage gauge**, = **cumulative raingauge**. A raingauge with a sump to catch and store rainwater over a long period, such as a month or year. There is usually some mechanism to prevent evaporation loss. Totalisers are used in areas of extremely difficult access.

total maximum daily load The estimated maximum pollutant load that a water body can receive without violating the relevant water quality standards. The TMDL is widely used in American water management.

total potential ... *of a soil* ... The sum of gravitational, pressure and osmotic head potentials.

Total retention storage The amount of water that must be added to the top of a snowpack before water is released from the base; in other words, before the snowpack is 'ripe'.

total runoff Combined surface runoff, interflow and groundwater outflow from a catchment.

total station The modern survey gear, which supersedes the old level and theodolite. At the push of a button a laser beam is returned from a distant reflector, and analysed to yield vertical and horizontal angles and distances. This is a very useful gadget for measuring hydraulic gradients, river cross-sections, irrigated areas *etc.* A total station saves a lot of time and effort—until the batteries run flat.

total storage ...*of a reservoir*... The combined active and dead storage.

total water potential The sum of gravitational and water potentials: $\phi = \Pi + P + Z$.

tower crib A tower extending from the base of a reservoir to its high water level, for drawing off water at different stages.

tower karst Residual limestone 'sugar loaf' hills. One of the last stages of karstic erosion from a previous to a current stable base-level of erosion. The type-area of tower karst is in around Guilin in Southern China. Marine inundated tower karst has also created the famously scenic Ha Long Bay in northern Vietnam.

towing tank A long trench-like tank over which a trolley can move at a precisely controlled rate. A current meter can be calibrated by dragging it through the water at a series of known rates. Note: a current meter *cannot* be accurately calibrated by holding it still in a stream of known discharge rate.

trace ...*of rainfall on weather records*... Rainfall that is known to have occurred, but which did not register in a raingauge, or the daily precipitation, p, of: $0 < p < 0.05$ mm. (<0.01 inches in America). Commonly denoted as 'tr' in rainfall records. In spreadsheet calculations the 'tr's may have to be replaced by '0.0'.

tracer   Any distinctive substance that can be used to quantitatively or qualitatively 'fingerprint' water. Tracers can be used to determine groundwater flow directions, discrete groundwater pathways, the mixing efficiency of two water sources or, by dilution gauging, surface water discharges. Tracers vary from ultra-sophisticated to the mundane and bizarre, and include: chopped straw, sawdust, coloured spores, non-pathogenic bacteria, lycopodium powder, marked live eels, common salt, fluorescent dyes such as fluorescein, uranine and rhodamine B or WT, aromatic compounds such as isobornyl acetate, and radioisotopes such as tritium, ^{24}Na, ^{42}K, ^{51}Cr, ^{82}Br, and ^{131}I. Some isotopic tracers can be detected at dilutions of 10^{-12}. The ideal tracer is cheap, harmless, chemically inert, mixes easily and is readily identifiable at low concentrations.

transition zone   = Thermocline.

transmissivity   (T) = **coefficient of transmissivity** = **transmissibility** *(archaic)*. A parameter indicating the ease of groundwater flow through a metre width of aquifer section (taken perpendicular to the direction of flow). Transmissivity and storativity are the two most important hydraulic properties of an aquifer. A typical range of values is about 10^{-1} to 10^4, in units of $m^2.day^{-1}$.

transmitted wave energy It is common for river banks and their related ecosystems to be damaged by the wake from boats.

This can be assessed upon the basis of transmitted wave energy, calculated by:

$$E_t = \sum_n \frac{\rho g^2 T_n^2 H_n^2}{64\pi}$$ where ρ is the water density,

g is gravitational constant, T is the peak-to-peak duration of each wave in the boat's wake, and H is each wave's amplitude. This gets complicated if there are wind waves as well as boat waves. Note that the wave period is as important as wave height. Most river banks are not well adapted to resist erosion from long-period waves.

transpiration   The process of water uptake through a root system, and subsequent loss to the atmosphere, by vegetation. Less than about 1% of a plant's water uptake is actually retained by the plant. The rest is lost by transpiration.

transpiration stress   A plant condition in which loss of moisture through the foliage exceeds the replenishment of moisture from the roots, resulting in partial wilting, yellowing or drying of leaves *etc.* Exposure to extreme desiccation may be diurnal and / or seasonal.

trapezoidal rule   An approximate formula for calculating areas (such as for stream cross-sections), by subdividing the area into n strips (n+1 ordinates), each of width b:

$$A = \int_a^b f(x).dx \approx \frac{b}{2}\{(y_0 + y_n) + 2(y_1 + y_2 + y_3 + \ldots + y_n)\}$$

trash marks   The debris on a river bank or riverside trees, buildings *etc.*, which indicate the high water mark of a floodwave. From this, it is possible to reconstruct the peak cross section of streamflow, and hence, by Manning's equation, the peak discharge rate.

trash rack = **trash screen**. A metal grid across a stream, culvert inlet, *etc.*, designed to filter out the worst of any flood débris. If not regularly cleared it can become blocked and cause more harm than good.

tree mould A feature of some Hawaiian type basalts, in which lava freezes around a tree, but the tree itself burns more slowly, to leave a cylindrical hole, thus providing hydraulic continuity between top and base of the lava flow.

tremie To place material, such as a gravel pack, to a particular point in a borehole by feeding it through a pipe.

trend surface analysis A variety of methods for applying a best-fit surface to a spatial array of data. Trend surfaces are rarely calculated to higher than third order.

triangular weir A triangular V-notched weir is designed to maintain accuracy of measurement at low discharge rates. The greater the angle of the 'V', the greater the range of flow that can be measured, but at the expense of accuracy. Calibration tables are readily available for the common angles, such as 30°, 45°, 60°, 90° and 120°. Calibration tables for the less common angles may have to be calculated.

triaxial diameter The mean particle diameter as measured along the major, minor and intermediate axes. It is tedious to determine, and hence seldom used in bed-load calculations.

trickle irrigation = **drip irrigation**, by tube direct to the root zone. It is expensive to install, and requires maintenance to prevent clogging by clay, algae *etc.*, but trickle irrigation is the most water-efficient method of irrigation.

tributyl tin 'TBT', a group of highly toxic compounds based upon the molecule $(C_4H_9)_3Sn$. It is mainly a hazard to marine organisms, but is also known to impact river and lake ecosystems. TBT has a half-life in seawater of about 6 hours, but a half-life in sediments of about 3.5 years. Anti-fouling paints leach about 5 µg TBT. $cm^{-2}.day^{-1}$. *See also* **organotin**.

tricone bit A drilling bit with three rotating toothed carbide cones. The most widely used of many rotary drilling bit designs.

trihalomethane (THMs). A group of carcinogenic chemicals, such as chloroform and bromoform, that are accidentally synthesised at very low concentrations by chlorination, or by flash desalination of water containing natural organics. The guidelines for drinking water vary from about 25 to 350 $mg.l^{-1}$.

trilinear diagram = **Piper diagram**. A pair of equilateral diagrams isometrically projected onto a common diamond-shaped field. The combination is a means of depicting the relevant proportions of cations and ions in a water analysis.

triple bottom line A notion that water projects, indeed development projects in general, must sign off on three basic criteria, namely: economic benefit, social acceptability and environmental sustainability. These are generally written into the ToRs of most donor or NGO-funded initiatives. Will climate change impact / vulnerability become a fourth bottom line?

triple porosity The complex porosity of karstic formations in which the rock contains inter-granular (primary), fracture (secondary), and open channel (tertiary) porosity. The effects can sometimes be seen as trimodal recession curves of karstic resurgences.

tritium (^{3}H). The radioisotope of hydrogen, with a half-life of 12.26 years. In the years following atmospheric nuclear tests (the 1950s and -60s), the greatly enhanced tritium concentration in meteoric water was a useful and approximately dateable tracer. Now however, its concentration has decayed or been diluted back to background concentrations, so it is no longer of much use.

tritium unit A working unit of tritium activity, equivalent to one tritium atom per 10^{18} atoms of water. 1 T.U. = 3.24 x 10^{-9} mCi.ml^{-1}. The natural background concentration of rainwater is typically about 1 or 2 TUs.

tropopause An atmospheric 'horizon' of minimum temperature, below which nearly all of the weather, moisture and atmospheric mass occurs. The height of the tropopause is variable, but typically occurs at about 12 to 15 km.

truncation error The error(s) incurred by discretizing a system; in simulating a system by a numerical grid, as opposed to a mathematically defined continuum.

trunk main The main pipe or pipes in a water reticulation system.

TUH Time unit hydrograph. The quickflow hydrograph resulting from 1 mm of effective rainfall, averaged over 'T' hours' duration. The TUH is the standardized unit response of a catchment, from which the response of any design storm can, in theory, be calculated. The intensity of a 'T hour' storm can be calculated as a multiple of the TUH, whilst the effects of prolonged rainfall are calculated by superposition of effects during successive time periods.

turbidity The opacity of water caused by suspended and colloidal particles, such as clay and organics.

turbine A rotodynamic machine for converting the kinetic energy of water to some other form of energy (usually electrical). There are two classes: *impulse* turbines, such as a Pelton wheel, are driven by a high velocity water jet at atmospheric pressure, whilst *reaction* turbines are enclosed by a 'scroll case', and convert a combination of kinetic and pressure energy to mechanical energy. *See also* **Kaplan** and **Francis turbines**.

turbine efficiency Energy supplied *by* the turbine, as a fraction of the hydraulic energy supplied *to* the turbine. It is typically about 0.9, or 90%.

turbine power $P_t = \eta_t \gamma Q h$, where η_t is the turbine efficiency, γ is the specific weight of water, Q is the flow rate and h is the head imparted.

turbulent flow Dynamic flow in which kinetic forces are much greater than viscous forces, and in which there is much internal shear stress. In turbulent flow, momentum is chaotically distributed.

turbulent velocity The velocity of water in a channel or pipe, above which the water will always be turbulent. At lower velocities, the water may be laminar or turbulent, depending on transience and conditions.

turnover Complete mixing of a water column in a lake which has been previously stratified.

turgor Rigidity of plant cells, and hence of the whole plant, due to high cell pressure. Loss of turgor produces limpness or wilting. Turgor is partly maintained by internal hydraulic pressure, and partly by osmosis.

type curves A means of graphical or semi-graphical solution to multivariate problems. *Examples:*
(1) a graph of electrode half spacing *vs.* apparent resistivity may be matched against a set of type curves for a 3 or 4 layer geological interpretation.
(2) a time - drawdown graph may be matched against type curves for a variety of different conditions and/or assumptions to solve aquifer tests. Several such type curves are available for different methods, including those of Papadopolos, Walton, Neuman, and Stretsova.

ultrafiltration Filtration of colloids, larger organics, smaller microbiota, *etc.* from a solution, as in reverse osmosis. This requires high pressures (or centrifugal force), and fine membranes such as some plastics, or gel-impregnated unglazed ceramic.

ultrasonic meter A streamflow measuring device that works by measuring the apparent velocities of contra-propagating beams of high frequency sound, diagonally across a stream.

ultraviolet light Electromagnetic radiation between the visible and X-ray part of the spectrum, between about 400 and 4 nm. It is divided into near-UV from 400 to 300 nm, far-UV between 300 and 200 nm, and extreme-UV of < 200 nm. UV has a strong bacteriocidal action, especially around 260 nm, and hence is sometimes used for water disinfection. Its chief drawback for this purpose is the limited exposure time in typical flow systems.

Disinfection requires about 20 mW.cm^{-2} and an exposure time of 15 to 30s. Less than 15% of clear sky insolation is in the UV region (at wavelengths of > 300 nm), but even a few hours of this UV exposure is enough to disinfect bacterially polluted water. UV absorbance at 206 nm can be used to measure nitrate concentrations.

unaccounted for water (= the 'angels' share'). The loss of water in a piped distribution system: taken as the difference between the pumped input from a wellfield or reservoir, and the metered supply (leakage + illegal connections). Domestic meters are typically calibrated to read 10 to 15% low. Hence much of the loss may be apparent rather than real. Zero loss is virtually impossible to achieve, but a modern well-maintained supply typically loses some 10 to 15%. Many old reticulation systems, in areas of rapid population growth, are hopelessly over-loaded, and lose up to about 50%.

unconfined conditions These occur in an aquifer in which the upper water surface is freely mobile and vertically unconstrained, with the surface at atmospheric pressure.

unconformity spring One of the most common spring-forming geological structures, in which an impervious lower boundary forces groundwater to the surface.

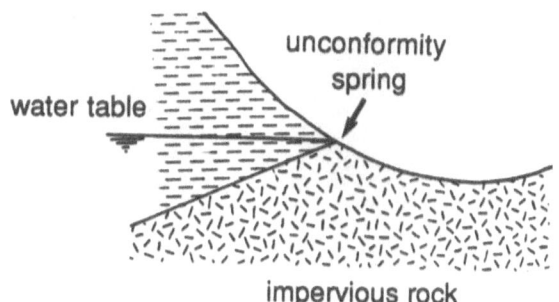

underfit stream A much smaller stream than would normally be expected from the catchment size; usually the result of tectonics, interception or changes in either climate or land-use.

underflow (1) Subhorizontal groundwater flow, generally associated with surface flow, as in the component of river flow that percolates through the streambed sediments.
(2) groundwater flow under a structure, such as an earth dam.
(3) streamflow beneath an ice cover.

undular jump A hydraulic jump with a smooth wave front, which occurs when the Froud number, Fr < 1.7.

uniform channel A channel of uniform cross section, slope and roughness characteristics, thus maintaining a constant velocity profile.

uniform flow Flow with the same velocity profile between upstream and downstream reference points; a state of constant velocity along a streamline.

uniformity coefficient A measure of the spread of particle sizes in a granular medium. Using a sieve analysis, the u.c. is the 40%, by weight, passing size ÷ the 90% passing size.

unit hydrograph = **unitgraph**, *see* **TUH** (time unit hydrograph).

universal time Greenwich Mean Time; the global reference time for all longitudes.

unsaturated zone = **zone of aeration**, = **vadose + soil water zones**. The upper part of an unconfined aquifer through which recharge taken place. The hydraulic conductivity of the unsaturated zone may be much greater than the saturated hydraulic conductivity of the same lithology. Due to capillary action, water is at less than atmospheric pressure in the unsaturated zone.

unsteady flow Changing stream discharge. The equations of unsteady flow involve momentum and acceleration, and are much more complex to solve than those of steady flow.

unsealed source Any radiating source which is free to interact with the environment, such as radioactive groundwater tracers.

unwatering Removal of surface water (as opposed to the removal of groundwater, which is 'dewatering'), as in the draining of rice paddies or coffer dams.

upconing In a stratified aquifer, especially a coastal aquifer with fresh water overlying sea water, upconing is the upward migration of the saline interface in hydrostatic compensation for a falling water table in and around a pumped well. *See also* the **Ghyben Harzberg relationship**.

upper critical velocity The velocity of a stream or conduit, at which Reynolds number, $R_N > 2800$, and in which flow is turbulent.

urban hydrology A sub-discipline of hydrology involving accelerated runoff and storm-water drainage from the largely impervious urban environment. The philosophy of urban hydrology is undergoing substantial change with the realization that wetlands and storm-water are potential resources, and not just wasteland and wastewater.

usage For water resources management purposes, consumptive water usage is commonly classified as irrigation or agricultural, industrial, municipal, domestic, and non-revenue water (leakage and illegal connections). WRM also has to account for non-consumptive water usage, mainly for recreational and hydro-power purposes.

useful storage = **active storage**.

USLE Universal Soil Loss Equation. An empirical equation for calculating the rate of fluvial soil erosion from rainfall and geomorphic parameters, simply defined as: A=RKLSCP, where A is the average annual soil loss (tonnes.km^{-2}.yr^{-1}), R is a rainfall erosivity factor, K is the soil erodability factor, L and S are physiographic factors, and C and P are crop management factors. Several authors prefer the modified version (*see* MUSLE) which uses runoff data in place of rainfall, although rainfall data is generally more available. There is some doubt about the applicability of the USLE in tropical environments. Decades of data are sometimes necessary to calibrate the USLE, so modern research is tending towards process analysis, and away from this 'black box empiricism'.

uvala A trough or closed valley with a rough bottom, containing karstic collapse features. Some authorities regard them as coalesced dolines in an advanced stage of erosion, but they may also be collapsed caverns.

vadose shaft A vertical or near-vertical shaft in the unsaturated zone of a limestone aquifer. It may have been initiated by collapse, or by pure dissolution, but has since been further excavated by falling water. The walls have rough irregular surfaces. The largest examples are hundreds of metres deep. *cf* **domepit**.

vadose zone = **zone of aeration**. The unsaturated zone comprising the capillary, intermediate (gravitational and pellicular), and soil water zones. (*Latin: vadosus* = shallow).

validation The process of tinkering with a computer model until the output (such as the configuration of a piezometric surface) matches the known historic data, ... and that is all! Validation does not imply that the model is necessarily 'valid' / correct / realistic. 'History matching' might be a more appropriate term than validation. *cf.* **post-audit**.

valley spring A spring caused by a a topographic depression (valley) locally dipping below a water table. They are normally found low in a valley, although perched water tables may produce a spring high up a valley side.

valve tower A structure within a reservoir, generally built close to a dam, with offtake valves at various depths.

vapour blanket The high humidity layer immediately above a lake surface.

vapour pressure The partial pressure of a vapour in the atmosphere. The conventional measure of its concentration.

variability index One of several statistical measures of spread of a dataset, defined as: $\dfrac{90\ percentile - 10\ percentile}{50\ percentile} \cdot 100\%$

See also **coefficient of variation**, **relative variability**, and **standard deviation**.

variable source model = dynamic catchment model. The Hewlett conceptual response of a catchment to rainfall, in which the surface drainage system undergoes ephemeral expansion, thereby increasing the efficiency of surface drainage.

variance The square of the standard deviation. A commonly used measure of the spread of data in a data set.

varves Annual sedimentary cycles of deposition in glacial or peri-glacial lakes. Each cycle consists of silty summer deposition and clay-rich winter deposition. These 'varved clays' or varved sediments carry such a reliable annual signature that they can be used for palaeoclimatic dating.

varzea Seasonal lakes, typically up to 10 metres deep, characteristic of the Amazon flood plain.

vauclusian spring A class of spring, unique to karstic areas, in which the discharge has the proportions of a river, such as 2 to 20 cumecs. Extreme discharges may even exceed 100 cumecs.

velocity head coefficient = energy coefficient, = Coriolis coefficient, 'α'. The velocity head of a stream $= \alpha \dfrac{v^2}{2g}$, (g = gravity and v = the mean velocity). In most stream or pipeflow calculations, α varies from 1.0 to 1.15; however in severely turbulent conditions, α may be significantly greater, such as 1.5 to 2.0.

velocity log See **sonic log**.

velocity of approach The mean water velocity in a channel or pipe immediately upstream of a flow measuring device.

Weirs, flumes *etc.* require a correction coefficient based upon the velocity of approach.

velocity profile (1) A contour map of water velocities across a stream section.
(2) A vector diagram of stream velocity with depth in the plane of flow. This is usually taken to be semi-parabolic in shape, but in turbulent high-friction streams, much more complex profiles occur, as shown.

velocity profiles under normal and rough conditions

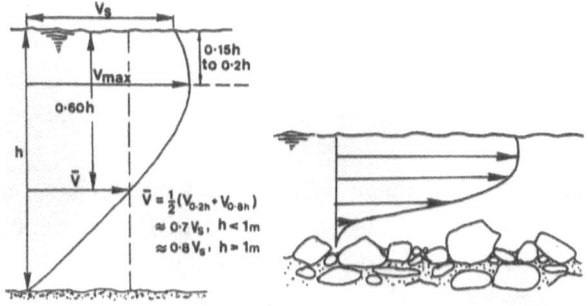

vena contracta The narrowest section of a jet, nappe or free-falling water.

ventilation Admission of air, at normal atmospheric pressure, to the under-side of a nappe. The purpose is to prevent the build-up of low pressure and cavitation.

ventilated chamber Apparatus for estimating transpiration on a small scale by enclosing the plant in a chamber and monitoring the humidity of inflow and outflow, together with other microclimatic parameters. The method was popular for a while but contains inherent inaccuracies. Other methods are now preferred.

venturi A laterally constricted streamflow device, such as a venturi flume. It measures discharge indirectly by measuring the differential pressure at constricted and non-constricted tappings.

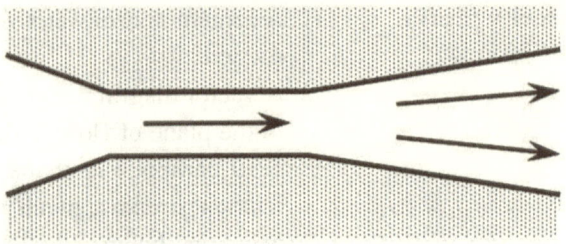

verkhovodka A peculiarly Russian term for water held in temporary near-surface saturated conditions.

vernier An accurate means of measuring instrumental distances, as in a hook gauge.

vertical wind shear Referring to a cumulus storm tower, differing wind velocities at different heights causes vertical wind shear. Consequently the storm's latent heat is spread over a larger footprint, whilst vertical convection cells are broken up. This breaks up and dissipates many storm cells, and may prevent a hurricane from forming in the first place. On the other hand, the creation of updraft rotation, and removal of precipitation through vertical wind shear, tends to prolong storms and may even enhance the storm's intensity.

vesicles Large pores in volcanic rock; in extreme cases accounting for 70 to 80 % porosity. However, this is non-linked porosity, and hence of no use for water storage or transmission.

Vibrio A family of aquatic pathogenic bacteria, including the notorious *V. cholerae*, which is responsible for some of the worst epidemics of water-borne disease.

virgin zone The aquifer volume, beyond some radial distance from a newly drilled borehole, in which the saturating fluid is entirely natural pore water (as opposed to the invaded zone).

virtual velocity (v_{hv}) is the average velocity of bed load particles. It does not discriminate between saltation, rolling, sliding or stationary intervals. *cf.* **actual velocity**. Basal stream velocity > v_h > (possibly >>) v_{hc}.

virtual water *See* **embedded water**.

viruses Hepatitis B is the most serious aquatic viral infection, although less than 1% of sufferers contract the disease through drinking water. As with other viral and some bacterial infections, a greater risk is via the ingestion of aquatic foods, notably shellfish. There are over 100 species of enteroviruses whose effects range from sub-clinical discomfort (common), to life-threatening conditions (rare). The specific virus responsible for a given illness is difficult to identify, but most can be controlled by water disinfection. Viruses are typically about a thousand times smaller than bacteria, in the size range of 10 to 400 nm. For example the Hepatitis B virion is 42 nm diameter, whilst the HIV virion, is about 180 nm diameter. In principle most viruses can be removed from water by means of nanofiltration using filter sizes of 20 to 50 nm. However, this requires a much longer filtration time than is practicable for public water supplies. Fortunately, aquatic viruses lose their infectivity with time.

viscosity (μ or η) Synonyms are: 'dynamic viscosity', 'absolute viscosity', 'coefficient of viscosity', and 'internal friction'. An intrinsic fluid property indicating the resistance to flow, or ease of shearing deformation. SI units are the 'Poiseulle' (PL), in $N.s.m^{-2}$, or kg.m⁻¹s⁻¹. *See also* **kinematic viscosity** and **dynamic viscosity**. The viscosity is slightly increased by salinity, and strongly decreased by both organic solutes, and temperature. For pure water at different temperatures, and for some other fluids, some values (PL x 10^{-3}) are:

water at 0 °C	1.80	air at 20°C	0.02
water at 10 °C	1.34	olive oil at 20°C	0.84
water at 20 °C	1.01	sea water at 20°C	1.01
water at 30 °C	0.80	ethyl alchohol at 20°C	1.20
water at 40 °C	0.65	50% sucrose at 20°C	100

viscous flow = **laminar flow** = **streamline flow**. Very even flow controlled by viscous forces. No turbulent mixing.

voids ratio $n = \dfrac{n}{1-n}$, where n is the porosity; it is the ratio of pore space to solid volume in a granular matrix. Extreme values range from zero to about 10 in a very loosely packed clay-rich sediment. For efficiently packed spheres of constant radius, e = 0.35.

volatilisation Vaporisation from a solid or liquid phase. Many dissolved organic pollutants volatilise into free air or into the unsaturated zone, with an equilibrium vapour concentration given by Henry's Law.

volume ...of a lake or river reach... may be approximated by either:
$$V = \sum_{i=1}^{n} \frac{l_c}{2}(A_i + A_{i+1}) \text{ or } V = \sum_{i=1}^{n} \frac{l_c}{3}\left(A_i + A_{i+1} + \sqrt{A_i . A_{i+1}}\right),$$
where l_c is the contour interval, and A_i is the plan area at successive depths.

vortex A whirlpool or circular eddy, usually with a vertical spin axis. In a *free* vortex, the tangential velocity increases towards the axis (as in water draining down a plug hole); in a *forced* vortex, the tangential velocity increases away from the axis (as in a stirred cup of tea).

vrtaca *See* **doline**.

vrulje A submarine freshwater spring discharging offshore from a karst aquifer.

wadi (Arabic) A normally dry valley or fluvial channel.

wading rods A set of metal tubes with end to end screw attachments, and to which a current meter is attached. Several facilities are often added such as cm-graduations, depth colour-coding, a heavy flat base for stand-alone stability, and an impellor axis indicator at the top for use in turbid water.

warm monomictic Describes a lake in which the epilimnion (surface stratum) cools during late winter, to the temperature of the hypolimnion, thereby causing instability and mixing. Warm monomictic lakes are found in Mediterranean and sub-tropical regions.

wash A small tributary gully, especially in arid regions.

WASH 'Water, Sanitation and Hygiene', is an international program to provide safe drinking water and sanitation through strategically targeted interventions that strengthen governance of the water and sanitation sectors. WASH is global in scope, but is most prominently active in poor and fragile regions, which are mainly concentrated in sub-Saharan Africa and southern Asia.

wash load　　　　Very finely suspended sediment, especially clay and organic material, which is transported by a stream without significant deposition.

wash out　　　　The scavenging of atmospheric dust and aerosols by rain-drops.

washed gravel　　Specifically, the gravel in a gravel pack. It is required to be well rounded, and washed clean of clay and organic matter.

water　　　　　The chemistry and physics of water may seem commonplace, but compared to other compounds, its properties, dominated by its quadropole molecular stucture, are quite extraordinary, and unlike any other known substance. Water mainly consists of the molecule $^1H_2^{16}O$ (on average 99.74%), but also includes $^1HD^{16}O$, $^1H_2^{18}O$, $^1HD^{18}O$, $D_2^{16}O$, $D_2^{18}O$, plus traces of ^{17}O equivalents.

water balance　　= **water budget**: said of a catchment or group of catchments. Any elaboration of the theme: Σ *inputs* = Σ *outputs* ± *any change in storage*. The water balance is the root concept from which sustainable water resource philosophies are derived.

water catch　　　An impervious slope or depression used for collecting rainwater.

water conditioning　Water treatment for organoleptic improvement, but exclusive of disinfection processes.

water conveyance efficiency . . . of an irrigation system. $\mu_c = 100 \cdot \dfrac{V_f}{V_t}$

　　　　　　　　where the percentage efficiency is the ratio of the volume of irrigation water actually reaching the crop, to the total amount of water diverted from source. *See also* **irrigation efficiency**.

water correction	Water treatment, other than filtration and chlorination, to prevent corrosion or encrustation.
water cushion	A pool of water at the toe of a dam, designed to dissipate the kinetic energy of overflows.
water equivalent	The water depth in mm, obtained from melting a snow sample. Caution: the water equivalent of snow is more spatially variable than that of rain, and is also strongly dependent upon the snow maturity.
water-hammer	The juddering effect of pressure surges back and forth along a length of pipe. This is typically caused by the sudden closing of a valve or tap. Serious damage can be caused by water- hammer, particularly if the pressure oscillates at resonant frequency.
water harvesting	Various methods of collecting and storing precipitation, usually on a small scale. Such techniques are becoming very important in marginal and arid lands. Techniques include: dew traps, roof gutters and tanks, sealed catchments, interception nets, micro- and macro-terracing.
water hyacinth	*Eichhornia crassipes*. A fast-spreading troublesome hydrophyte, originally from the Amazon basin and now across all the tropics, but most typically occurring in tropical and sub-tropical Africa, especially in the Congo and Nile basins. It is notorious for clogging river and lake surfaces.
watering in	The act of adding water to electrode points prior to an electrical resistivity traverse. In dry desert conditions the high surface resistance may require electrode points to be watered, possibly with the addition of salt, several times during the week prior to the traverse. It is laborious and tedious, but may make the difference between success and failure.

waterlogged Totally saturated condition.

water of constitution = **water of crystallisation**. Chemically bound water, as in the mineral gypsum: $CaSO_4.2H_2O$. Most sulphates incorporate some water, and some can exist in several hydration states. At high temperatures mirabilite, $Na_2SO_4.10H_2O$, even has the capacity to dissolve in its own water of crystallisation.

water potential = **total suction** = **retention potential**. In soil or plants, the sum of osmotic and pressure or matric potentials; ie $\phi = \Pi + P$. The combined physical and chemical suction which a plant exerts to draw up moisture. The soil water potential is related to the *relative vapour pressure* by: $\Pi + P = \dfrac{\rho RT}{M} \ln \dfrac{e}{e_0}$ where ρ is the water density, R is the gas constant per mole, T is in °K, and M is the molecular weight of water. $\dfrac{e}{e_0}$ is the ratio of actual vapour pressure to the vapour pressure of free water at the same temperature.

water-related diseases These are classified as:
i) *water-based diseases;* caused by pathogens whose life cycle involves an aquatic host such as a mosquito or snail,
ii) *water-borne diseases;* caused by drinking pathogen- contaminated water, and
iii) *water-washed diseases;* caused by non-ingested infection. Examples are i) schistosomiasis and malaria, ii) typhoid and cholera, and iii) scabies and amoebic meningitis.

water rights Water rights most commonly relate to the allocation of scarce irrigation water. In the wider sense, water rights cover the legal basis for water allocation, ownership, development, access, prioritization, protection, and conservation. The subject is immensely complicated, and varies massively according to culture, country, state, and local authority.

water rights cont. In many countries, and in nearly all developing countries, the law governing water rights is outdated, poorly defined, incomplete and inconsistent (between laws covering different sectors). Lawyers with a sound technical understanding of the water sector are a rarity, yet there is growing scope for lawyers to get even richer over disputed water rights.

watershed (English):The *boundary* of a drainage basin, or drainage divide. (American): The *area* of a drainage basin.

water softening In order to prevent scaling, water may be treated to remove Ca and Mg alkalinity. The two methods are:
(1) Precipitation of these salts as carbonate by the addition of hydroxides, followed by filtering, and
(2) ion exchange on synthetic resins or zeolites to replace the divalent ions with sodium.

water table = **the 'phreatic' or 'free' surface**: The saturated-unsaturated interface within the ground. Also known as the free atmospheric pressure surface.

water-trading curve An optimum trade-off between urban and irrigation water usage can be represented by the amount and price of acceptable trade, Q_e and P_e, respectively on the graph below.

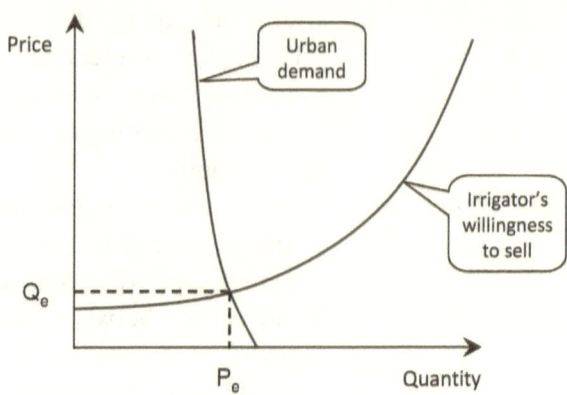

water witching	See **dowsing**.
water year	It is often convenient for surface water flow or groundwater levels to be accounted from a seasonal minimum rather than from January 1st. This minimum is the start of the water year, which varies with region and hemisphere. In the USA the water year nominally starts on October 1st.
WDSC	Water Dispersible Sand and Clay; an index of soil stability defined as: $$WDSC = \frac{\text{water dispersible solids of} < 60\,\mu m}{\text{total solids of} < 60\,\mu m} \cdot 100\%$$
weathering	A 'bucket term' for the sum of many physical and chemical processes involved in the *in situ* breakdown of rocks. Climate, rather than rock-type, is the main controlling influence upon rates and processes of weathering. As a first approximation, the wetter the climate, the faster the weathering.

Weber number A dimensionless ratio of inertial to surface tension forces, given by: $W_e = \dfrac{\rho v^2 L}{\sigma}$ where ρ is the density, v is the velocity of flow, L is a characteristic length, and σ is the surface energy (tension). W_e" is used in considerations of drop formation, capillarity and in vorticity. It is conceptually analogous to Reynolds number.

Weibull $AEP = \dfrac{m}{N+1}$. One of several plotting position formulae for estimating average exceedance probabilities. Some people prefer the **Gringorten estimate** *(q.v.)*, although there is no objectively best formula.

weighing lysimeter A large and complex form of field lysimeter in which an entire portion of rock, complete soil section and typical vegetation are isolated in a weighing tank. It is difficult and expensive to set up, but is capable of yielding excellent data on the gross water balance of the environment in question.

weir height The vertical distance from the upstream bed to the weir's crest.

weirs and flumes Structures for establishing artificial section control, whereby discharge is uniquely related to stage. A huge variety of two- and three-dimensional designs is available. A weir is an overflow structure, whilst a flume creates lateral constriction.

well capacity *See* **specific capacity**.

well completion diagram A schematic diagram showing all the well construction details, such as depths of screen, connections, telescoping, cementing, gravel pack, *etc.*

well efficiency The theoretical drawdown in a well, as a percentage of the observed drawdown. Well efficiency is measured by step-drawdown tests. *See* **excess drawdown**.

wellfield A cluster of wells all drawing upon a common aquifer system, such as for a town water supply. For economy of pipework and logistics, the wells should be closely spaced. However, if the wells are too close, their cones of depression mutually interfere, causing excessive drawdown. Too few wells, low transmissivities, inadequate recharge or extended pumping may also result in excessive drawdown. Good wellfield design and operation is therefore a 'tricky compromise' between all these factors. The modern wellfield manager has to juggle such issues as groundwater sustainability, water quality, reliability of supply, pumping and maintenance costs, and environmental protection. Wellfield protection, to maintain groundwater quality is becoming ever more important, particularly as urban sprawl encroaches upon the active area of the wellfield.

well function = the **Theis well function**, 'W(u)', also expressed as -Ei(-u). By making a simplifying assumption or two, the well function can be utilised to analyse non-equilibrium aquifer test data for 'S and T'. It can be expanded into the infinite series:

$$W(u) = \int_u^\infty \frac{e^{-y}dy}{y} = -0.5772 - \ln u + u - \frac{u^2}{2.2!} + \frac{u^3}{3.3!} - \frac{u^4}{4.4!} + \ldots$$

Where $u = \dfrac{r^2 S}{4Tt}$, r is the radius from observation well to pumped well, S is storativity, T is transmissivity, and t is the time since the start of pumping.

well logging	= **borehole logging** = **geophysical well logging** = **mud logging** ≈ down hole geophysics. Probes are available for sensing about 30 different parameters down a well, but most well logging only records profiles of a few parameters, e.g. S.P., long and short resistivity, γ, caliper and neutron logs.
well loss	That part of the drawdown in a well which is due to flow within the well screen and gravel pack (as opposed to the head loss in the aquifer formation). It is often approximated as the non-Darcian or turbulent head loss, and can be estimated by a step drawdown test. *See* **well efficiency**.
well point	A simple form of shallow well consisting of a strong perforated screen with a sharp point, capable of being driven into soft soil or alluvium. Well points are widely used to drain mine tailings, and construction sites, or to develop shallow water supplies.
well screen	*See* **screen**.
well shooting	= **explosive fracturing**, = **explosive development** *(q.v)*.
well storage	In the case of large diameter bores or dug wells the water in well storage can be a significant part of the overall yield. For aquifer tests the early results are influenced more by well storage than by aquifer characteristics. The rule of thumb is: the drawdown of an aquifer test, to time 't' is irrelevant up to $t = \dfrac{250r^2}{T}$ where r is the well radius and T is the transmissivity. Also, stagnant well-water exposed to air may be chemically unrepresentative of formation hydrochemistry, so the convention is to pump 2.5

to 3 times the well storage before sampling for analysis.

well vent A small hole or air valve in the surface casing of a bore designed to maintain atmospheric pressure in an otherwise sealed well. This is particularly necessary for instrumental water level monitoring.

well yield The maximum sustainable pumped volume in $m^3.day^{-1}$.

Wenner array In shallow resistivity surveying, the Wenner array is a configuration of four equidistant electrodes comprising two inner potential electrodes, and two outer current electrodes.

wetland A frequently flooded or perennially wet, if not saturated marshland, bog, fen or swamp. For many years wetlands were regarded as wastelands 'to be drained for something more productive'. They are now recognized as being vital to surface water quality and wildlife. In particular, wetlands are the natural bio-physical water filters and flow regulators of natural ecosystems.

wet-line correction A correction required to estimate the true depth of a suspended current meter. *See* **air-line correction**.

wetting angle = Angle of contact, θ_w. The angle between a solid surface, and the interface between two immiscible fluids.

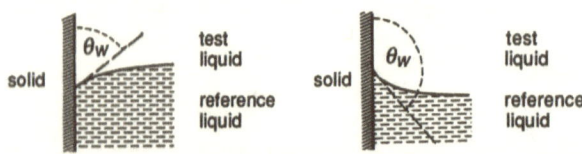

$\theta_w < 90°$: test liquid is wetting $\theta_w > 90°$: test liquid is non-wetting

Dictionary of Hydrology and Water Resources

wetting front   The leading edge of a saturated zone that moves downward through a soil or unconfined sediment during and after a rainstorm or after irrigation. Perhaps counter-intuitively, some 12 to 24 hours after rainfall/irrigation the wetting front seeps deeper into the soil beneath clumps of vegetation rather than beneath bare soil.

wetted perimeter   For a stream section perpendicular to flow, it is the perimeter length of the stream bed, *not* including the surface width.

whitening   Backscattering of white light by suspended microcrystalline carbonate minerals, such as calcite and aragonite. This is a transient feature of some lakes which undergo seasonal carbonate precipitation. Lakes which exhibit whitening include Lake Michigan (USA), Blue Lake, and the Coorong (both in South Australia).

white water   (1) Turbid rivers containing a high concentration of suspended sediments, usually derived from erosion in mountainous source areas, such as the headwaters of the Amazon, as distinct from the blackwater tributaries from the lowlands.
(2) Turbulent rivers with numerous alternations of supercritical and subcritical flow, thereby providing sport for canoeists, rafters, *etc.*

width of contribution   The plan width of groundwater flow which contributes to the yield of a single pumped well.

wild flooding   A very wasteful and haphazard method of irrigation in which channels are allowed to overspill, resulting in random overland flow.

willingness to pay   In many countries of low income and decreasing per capita water availability the 'willingness to pay' is an important and growing issue. There is a trade-off between level of supply, and affordability. In many countries this is complicated by the differential between the water utilities' promised and delivered benefits, and by the perceived lack of infrastructure maintenance. 'Willingness to pay' should be an objective social analysis, but it is frequently a socio-political circus.

wilting point The soil moisture content at which permanent wilting of plants occurs; the volumetric water content of the soil at the limit of plant extraction.

wind shield Protective devices designed to reduce micro-aerodynamic turbulence around the rim of a precipitation gauge. There are various designs, such as Nipher or Tretyakov shields. None of these are entirely successful, so always be aware of potential under-catch errors in wind-exposed locations.

windermere A 'unit' of water volume sometimes used by the popular press. Under average conditions the volume of Lake Windermere, in the English Lake District, is about 315 million cubic metres, = 1.54 mid-tide Sydney Harbours, 126,000 Olympic swimming pools, 0.00019 Lake Ontarios, or about 28 million elephants (*Loxodonta Africana,* sans their 160 litre drinking capacity).

winter irrigation (cool climates only) Irrigation during the non-growing season to bring the soil to field capacity at the onset of germination.

wire-wound Also known as **wedge wire**. A highly efficient screen comprising a stainless-steel wire continuously wound around, and welded to, longitudinal stay wires. The cross-section of the wire is designed to minimize particle clogging. It is expensive compared to slotted or perforated screen, but gives higher yields and lower 'well losses'.

World Meteoric Line Also known as the **Meteoric water line**, defined as: $\delta D = 8\delta^{18}O + 10$ although there are several regional deviations from this line, especially in warmer climates. It is a line of best fit, which shows the close correlation between *mean* δD and $\delta^{18}O$ of precipitation throughout the world.

xeriscape   Dry landscaping: the use of dryland water resources in an appropriate and conservative manner. This is particularly important in desert urban and suburban environments. Aspects include: minimizing lawn areas, the use of mulches, shade trees and drought tolerant plants, efficient irrigation, water harvesting and recycling.

xerophyte   A class of vegetation which has adapted to arid conditions by means of an extensive but shallow root system to absorb surface moisture (such as morning dew). They have complex systems of osmotic adjustment. *cf.* **phreatophytes**. Xero-halophytes are desert species of halophytes which often secrete salt, and which typically occur as low biodiversity ecosystems.

xylem   Part of the root-stem-leaf continuum in which water is transported along nutrient flowpaths, primarily to facilitate transpiration from the leaves. In trees, sapwood consists of active xylems, whilst in heartwood, the xylems have been blocked by the deposition of lignin.

Y

yield point   The threshold pressure that a pump requires to initiate the circulation of thixotropic drilling fluid.

youthful river   A river in an early stage of development, with a steep gradient, much turbulence and rapid erosion. Most mountain streams and headwaters are youthful.

Z

zagira   (Arabic) An old-fashioned type of dug well which is bailed by an animal walking up and down a ramp.

zero defect   A Japanese concept that 'it is always cheaper to do it right the first time'. It applies to almost anything, but is particularly the case with well drilling. Trying to correct a badly drilled or poorly cased borehole is inordinately time-consuming, difficult, frustrating and expensive.

zero flux plane   A notional horizontal plane across which, there is no net vertical moisture movement. The most common case is that of a 'recharge pulse', below which there is a downward (gravitational) moisture flux towards the water table, and above which there is an upward soil moisture flux caused by evapotranspiration. Convergent and multiple zfps can occur within the same vertical profile. zfps also develop within slowly dewatering aquitards which have been depressurised above and below.

zero point of charge pH_0 or pH_{zpc}. The pH of a colloidal surface at which there is neither a net + or net - charge. Hence there is no electrostatic adsorption. $pH_{zpc} \approx 2$ for clays and quartz, 7 for Fe-oxides / hydroxides, or 9 for Al-oxides / hydroxides.

zoned dam An earth dam built in sections of different material, with each section having different permeability or density. For example, an 'impervious' upstream face with a permeable toe to facilitate the release of hydrostatic pressure.

zone of influence The area around a pumped well, tile drain, quarry, foundation *etc.*, in which there is detectable drawdown. In the case of radial groundwater flow to a well, this extends to the *radius* of influence.

zoogleal layer A thin bio-gel layer, mainly of aerobic bacteria, which forms on the surface of sand filters, especially in sewage works. It is biologically very active, and is of great importance in secondary wastewater treatment.

z/r relationship An empirical relationship between radar reflectivity and rainfall. *See* **radar equation**.

ACRONYMS AND ABBREVIATIONS COMMONLY USED IN THE CONTEXT OF HYDROLOGY AND WATER RESOURCES

NB. Upper case is normally used, but where common usage is almost invariably lower case, it is left as such in this listing. Otherwise, no particular significance is attached to either case.

AA	Activated Alumina
AAS	Atomic Absorption Spectrometry
ABS	(1) Acrylonitrile Butadiene Styrene, (2) Alkyl Benzene Sulphonate
AC	Activated Carbon, see GAC
ACSAD	Arab Centre for the Studies of Arid Zones and Dry Lands
ACWR	Australian Centre for Water Resources
ADB	Asian Development Bank
AE	Actual Evaporation
AEP	Annual Exceedance Probability
AES	Annual Exceedance Series
AF	Application Factor
AfDB	African Development Bank
AGU	American Geophysical Union
AHD	Aswan High Dam
ALARA	As Low As Reasonably Achievable (referring to radiation)
AMC	Antecedent Moisture Condition
AMD	Acidic Mine Drainage, or Acid and Metalliferous Drainage
AMRA	American Water Resources Association
AMS	(1) Accelerator Mass Spectrometry, (2) Annual Maximum Series
AMU(N)	Atomic Mass Unit (Number)
ANC	Acid Neutralizing Capacity
ANN	Artificial Neural Network
AOGCM	Atmospheric-Ocean (coupled) Global Climatic Model
AOP	Advanced Oxidation Process

APDC	Ammonium Pyrolidine-Dithiocarbamate
APHA	American Public Health Association
API	(1) American Petroleum Institute, (2) Antecedent Precipitation Index
APWF	Asia Pacific Water Forum
AR	(1) Analytical (grade) Reagent, (2) Activity Ratio, eg. of 238U/234U
ARC	Annual/Assured Refill Curve
ARF	Areal Reduction Factor
ARI	Average Recurrence Interval
ARIMA	Auto-Regressive Integrated Moving Average
ARMA	Auto-Regressive Moving Average
ASCE	American Society of Civil Engineers
ASR	Aquifer Storage and Recovery
ASV	Anode Stripping Voltametry
AVHRR	Advanced Very High Resolution Radiometer
AWRC	Australian Water Resources Council
AWS	Automatic Weather Station
AWSC	Available Water Storage Capacity
AWWA	(1) American Water Works Association (2) Australian Water and Wastewater Association
BIS	Bayesian Inference System
BGH	Bacterial Growth Efficiency
BHS	British Hydrological Society
BHT	Bottom Hole Temperature
BMP	Best Management Practice
BMR	Bureau of Mineral Resources (Austr.)
BN	Bayesian Network
BOD	Biological Oxygen Deficiency ≈ Biochemical Oxygen Demand
BP	(years) Before Present
BS	British Standard(s)
BTEX	Benzene, Toluene, Ethylbenzene and Xylene
CAD	Computer Aided Design
CAE	Carbon Alcohol Extract
CAM	Crassulacean Acid Metabolism

CAS	Complex Adaptive System
CAZRI	Central Arid Zone Research Institute (India)
CC	Climate Change
CCD	Cold Cloud Duration
CCE	Carbon Chloroform Extract
CDF	Cumulative Density Function
CDTA	trans-1-, 2-diaminocyclohexane-NNN'N'-tetra-acetic acid
CEC	(1) Cation Exchange Capacity (2) Compounds of Emerging Concern
CFSTR	Continuous Flow Stirred Tank Reactor
CGIAR	Consultative Group on International Agricultural Research
CGS	Centre for Groundwater Studies (Australia)
CHC	Chlorinated Hydrocarbon
CI	Collaborative Institutions
CIDA	Canadian International Development Agency
CLSA	Closed Loop Stripping Analysis
COD	Chemical Oxygen Deficiency (or Demand)
CoP	Chief of Party (American equivalent of TL)
CP	Cathodic Protection
CPE	Chlorinated PolyEthylene
CPS(M)	Counts Per Second (Minute)
CPT	Conditional Probability Table(s)
CRREL	Cold Regions Research and Engineering Lab. (USA, includes SIPRE)
CSPE	Chloro-Sulphonated PolyEthylene
CTM	Cohesive Transport Model
CTT	Cloud-Top Temperatures
CWA	Clean Water Act (USA)
DAD	Depth-Area-Duration (rainfall curves)
DALR	Dry Adiabatic Lapse Rate
DANIDA	Danish International Development Agency
DBA	Discrepancy Based Analysis
DBP	Disinfection By-products
DDAF	Depth-Duration-Area-Frequency curves
DDE	Dichloro-diphenyl-dichloroethylene
DDT	Dichloro-diphenyl-trichloroethane

DEM	Digital Elevation Model
DFID	Department for International Development (UK)
DMA	District Metered Area
DMI	Domestic, Municipal and Industrial (water usage)
DNAPL	Dense Non-Aqueous Phase Liquid (groundwater pollutant)
DO	Dissolved Oxygen
DOC	Dissolved Organic Carbon
DOP	Dissolved Organic Phosphorous
dpc	drill-pipe connection
dpm	Disintegrations per minute
DRI	(1) Desert Research Institute (USA)
	(2) Drought Risk Index
DRRD	Dual-wall Reverse circulation Rotary Drilling
DSS	Decision Support System
DTA	Differential Thermal Analysis
DTG	Differential Thermo-Gravimetry (= TGA)
DTM	Digital Terrain Model
DWFM	Digital Water Flow Meter
DWL	Dynamic Water Level
EA	(1) Executing Agency
	(2) Environmental Assessment
	(3) Evolutionary Algorithm (or Evolutionary Strategy)
EAP	Emergency Action Plan (generally for floods)
EBRD	European Bank for Reconstruction and Development (including water projects)
EC	Electrical Conductivity / Conductance
EDF	European Development Fund
EDTA	Ethylenediamenetetra-acetic acid
EDXRF	Energy Dispersive X-Ray Fluorescence
EE&A	Environmental Education and Awareness
EEC	European Economic Community
EEM	Ether Extractable Material
EGL	Energy Grade Line
EIA	Environmental Impact Assessment
EIRR	(Overall) Economic Internal Rate of Return
EIS	Environmental Impact statement
ELA	Equilibrium Line Altitude (of glaciers)

Dictionary of Hydrology and Water Resources

EM	Electromagnetic (conductivity)
EPA	Environmental Protection Agency
EPE	Equivalent Porous Medium
EQO(S)	Environmental Quality Objective (Standard)
ERCE	European Regional Centre for Ecohydrology
ERS	Event Recognition System
ESD	Environmentally Sustainable Development
ESP	Extended Streamflow Prediction
EU	European Union
EV1,2,3	Extreme Value Distributions (types 1, 2, and 3)
EWARS	Early Warning Alert and Response System
FAM	Floating Aquatic Macrophyte
FAO	(United Nations) Food and Agriculture Organization
FCR	Free Chlorine Residual
FDM	Finite Difference Model
FEM	(1) Fixed frequency Electro-Magnetic
	(2) Finite Element Model
Fm	Formation (geological)
FO	Forward Osmosis
FRE	Fractured Rock Environment
FS(R)	Feasibility Study (Report)
GAB	Great Artesian Basin
GAC	Granular Activated Carbon
GAP	Gender Action Plan
GCM	Global Climatic Model ≈ General Circulation Model
GCMS	Gas Chromatograph Mass Spectrometer
GEF	Global Environment Facility
GEMS	Global Environmental Monitoring Station/System
GEV	General Extreme Value distribution
GEWEX	Global Energy and Water cycle EXperiment
GGE(e)	Greenhouse Gas Emissions (CO_2 equivalent)
GGn	Generalized Gamma distribution. (n = 2, 3 or 4 parameters)
GHGe	Green House Gas (including the CO_2 equivalent of CH_4, N_2O, etc.)
GI	Galvanized Iron
GIGO	Garbage In—Garbage Out (golden rule of computing)

GIS	Geographic Information System
GISP	Greenland Ice Sheet Precipitation
GOES	Geostationary Orbiting Environmental Satellite
GPR	(1) Ground-Penetrating Radar
	(2) General Purpose Reagent
GPS	Global Positioning System
GPU	Graphics Processing Unit
GRP	Glass Reinforced Plastic (as in well casing)
GSC	Gross Storage Capacity
GSMS	Gas Source Mass Spectrometer
GTZ	German Technical Cooperative Agency
GWP	Global Water Partnership
HBI	Hilsenhoff Biotic Index (US, based upon macro-invertebrates)
HCB	Hexachlorobenzene
HDI	Human Development Index
HDL(C)	Highest Desirable Limit (Concentration)
HDPE	High Density PolyEthylene
HEP	Hydro-Electric Power
HGL	Hydraulic Grade Line
HH	Household
HPC	Horizontal Plane of Collimation
HPI	Human Poverty Index
HPL(C)	Highest Permissible Limit (Concentration)
HPLC	High Performance Liquid Chromatography
HRI	High Risk Intervention
HTME(D)	Horizontal Tube Multi-Effect (Distillation)
HWS	Hybrid Water System
I&D	Irrigation and Drainage
I/O	Input/Output
IA	Implementing Agency
IAEA	International Atomic Energy Agency
IAH	International Association of Hydrogeologists
IAHR	International Association of Hydraulic Research
IAP	Ion Activity Product
IASH	International Association of Scientific Hydrologists
IBI	Index of Biotic Intensity

Dictionary of Hydrology and Water Resources

IBRD	International Bank for Reconstruction and Development (= World Bank)
IC	Ion Chromatography
ICARDA	International Center for Agricultural Research In Dry Areas
ICE	Institution of Civil Engineers (U.K.)
ICEWaRM	International Centre of Excellence in Water Resources Management
ICHARM	International Centre for Water Hazard and Risk Management (Japan)
ICM	Integrated Catchment Management (≈ TCM)
ICP	International Co-operative Programme
ICP-AES	Inductively Coupled Plasma-Atomic Emission Spectrometry
ICP-MS	Inductively Coupled Plasma-Mass Spectrometry
ID	Inside Diameter
IDOD	Immediate Dissolved Oxygen Demand
IE	Irrigation Efficiency
IFD	Intensity—Frequency—Duration (graph)
IHD	International Hydrological Decade (1965-1974)
IHE	Institute of Water Education (Delft, Netherlands, UNESCO-based)
IHP	International Hydrological Programme
ILRI	International Institute for Land Reclamation and Improvement (Netherlands)
IMF	International Monetary Fund
IMWA	International Mine Water Association
INAA	Instrumental Neutron Activation Analysis
IOD	Immediate Oxygen Demand
IoH	Institute of Hydrology (U.K., now defunct, merged with ITE)
IOS	Immobile Oil Saturation
IPCC	Inter-governmental Panel on Climate Change
IR	Inception Report
IRRI	International Rice Research Institute, q.v.
ISE	Ion Selective Electrode (Electrochemistry)
ISO	International Standards Organization
IT	Information Technology
ITCZ	Inter-Tropical Convergence Zone

ITE	Institute of Terrestrial Ecology (UK)
IUH	Instantaneous Unit Hydrograph
IUPAC	International Union of Pure and Applied Chemistry
IWHR	Institute of Water Resources and Hydropower Research (China)
IWIC	International Waste Identification Code
IWM	Integrated Waste Management
IWMI	International Water Management Institute
IWRM	Integrated Water Resources Management
IWSA	International Water Supply Association
JCPDS	Joint Committee on Powder Diffraction Standards (for X-ray diffraction)
JICA	Japanese International Cooperation Agency
JMP	Joint Multipurpose Project
JTU	Jackson Turbidity Unit
JTWC	Joint Typhoon Warning Center
LAI	Leaf Area Index
LC(D)	Lethal Concentration (Dose)
LDC	Least Developed Countries
LGU	Local Government Unit
LNAPL	Light Non-Aqueous Phase Liquid
LOI	Loss on Ignition
LOS	Line Of Sight
LSI	Large Scale Irrigation
LTME(D)	Low Temperature Multi-Effect (Distillation)
LUST	Leaking Underground Storage Tank (re-think that one, guys!)
LWM	Lower Water Mass (of a stratified water body)
LWP	Leaf Water Potential
M&E	Monitoring and Evaluation
MA(T)C	Maximum Acceptable (Toxic) Concentration, $\approx$ HPL
mamsl	Meters above mean sea level
MBA	Master of Business Administration (and some less polite variants!)
MCA	Multi-Criteria Analysis

MCI	Multiple Cropping Index
MCL(G)	Maximum Contaminant Level (Goal)
MCRT	Mean Cell Residence Time
MCU	Micro- Controller Unit
MDBA/C	Murray-Darling Basin Administration/Commission (Austr.)
MDG	Millennium Development Goals (of UN agencies)
MDTOC	Minimum Detectable Threshold Odour Concentration
MED	Multiple Effect Desalination (see also LTME and HTME)
MFI	(1) Membrane Filtration Index
	(2) Modified Fouling Index
MIB	2-methyl-isoborneol
MIBK	Methyl-iso-butyl ketone
MIS	Management Information System
ML	Maximum Likelihood
MM	Method of moments
MOSS	Minimum Operating Security Standards
MoU	Memorandum of Understanding
MPC	Maximum Permissible Concentration
MPN	Most Probable Number (of bacteria per 100 mls, as in microbiological assay)
MRC	Mekong River Commission
MRS	Magnetic Resonance Sounding
MSF	Multi-Stage Flash (distillation)
MSFL	Micro-Spherically Focused Log
MSS	Multi-Spectral Scanner (or Multispectral Scanning System)
MUSLE	Modified Universal Soil Loss Equation
NARBO	Network of Asian River Basin Organizations
NAPL	Non-Aqueous Phase Liquid
NAS	National Academy of Science (U.S.A.)
NASQAN	National Stream Quality Accounting Network (USA)
NAWDEX	National Water Data EXchange (USA)
NBI	Nile Basin Initiative
NBOD	Nitrogenous Biochemical Oxygen Deficiency
NBS	National Bureau of Standards (U.S.A.)
NCCCD	National Committee for Climate Change and Desertification
NERC	Natural Environment Research Council (U.K.)
NGO	Non-Government Organization

NMR(I)	Nuclear Magnetic Resonance (Imaging)
NOAA	National Oceanographic and Atmospheric Administration (U.S.A.)
NOAEL	No Observed Adverse Effect Level (i.e. chemical concentration)
NOM	Natural Organic Matter (soluble)
NRC	Nuclear Regulatory Commission (USA)
NRPB	National Radiological Protection Board (UK)
NRW	Non-Revenue Water
NSF	National Science Foundation (U.S.A.)
NTA	Nitrilotriacetate
ntp	normal temperature and pressure (obsolete: see STP)
NTU	Nephelometric Turbidity Unit
NWWA	National Waste Water Association (USA)
O&M	Operation and Maintenance
OD	(1) Ordnance Datum (= Newlyn mean sea level, U.K.), (2) Outside Diameter
OECD	Organization for Economic Cooperation and Development
OHP	Operational Hydrology Programme (of W.M.O.)
OLS	Ordinary Least Squares (model)
ORC	Oil Retention Capacity
ORSTOM	Office de la Recherche Scientific et Technique Outre-Mer (Fr.)
PAC	Powdered Activated Carbon
PAH	Polycyclic Aromatic Hydrocarbon(s)
PAM	(or PAE) Primary Amoebic Meningoencephalitis
PCA	Principle Component Analysis
PCB	Polychlorinated Biphenyl(s)
PDB	Pee-Dee Belemnite (The primary 13C isotopic standard)
pde	partial differential equation
PDF	Probability Density Function
PDSI	Palmer Drought Severity Index
PE	Potential Evapotranspiration
PFU	Plaque Forming Units
PLF	project logical framework
pm	picometer ($1Å = 100pm$)

pmc	percent modern carbon
PMF	Probable Maximum Flood
PMP	Probable Maximum Precipitation
PMU(O)	Project Management Unit (Office)
POC	Particulate Organic Carbon
POPs	Persistent Organic Pollutants
POT	Peaks Over Threshold (time series)
POW	Productivity of Water
ppb	Parts per billion (1 in 10^9)
ppm	Parts per million (1 in 10^6)
PPP	Public-Private Participation
ppt	Parts per trillion (1 in 10^{12})
PPTA	Project Preparation Technical Assistance
PVC	Poly-Vinyl Chloride
PWM	Probability Weighted Moment (statistical theory)
PWP	(1) Pore Water Pressure, (2) Permanent Wilting Point
QA/QC	Quality Assurance / Quality Control
QCBS	Quality and Cost-Based Selection
RAFT	Rising Air Flotation Technique (for streamflow measurement)
RCM	Regional Climate Model
RDI	Recommended Daily Intake
REDD	Reducing (GHG) Emissions from Deforestation and Forest Degradation
RH	Relative Humidity
RIP	Regional Indicative Programme
RMSE	Root Mean Square Error
RO	Reverse Osmosis (if sea water or brackish water: SWRO, BWRO)
RoP	Rate of Penetration (of a drill bit)
RP	Return Period (obsolete, see ARI)
RSI	Ryznar Stability Index
RWL	Recovering Water Level
SA	Social Assessment

SALR	Saturated Adiabatic Lapse Rate
SAR	(1) Sodium Adsorbtion Ratio,
	(2) Synthetic Aperture Radar
SCADA	Supervisory Control and Data Acquisition (q.v.)
SCC	Surface Contact Cover
SCOPE	Scientific Committee on Problems of the Environment
SCSA	Soil Conservation Society of America
SEA	Strategic Environmental Assessment
SDI	Sludge Density Index
SDR	(1) Standard Dimension Ratio
	(2) Spatial Decay Ratio
SDWA	Safe Drinking Water Act (USA)
SEM	Scanning Electron Microscope (microprobe)
SI	(1) Systeme Internationalle,
	(2) Saturation Index
SIDA	Swedish International Development Agency
SIP	Strongly Implicit Procedure
SIPRE	Snow, Ice and Permafrost Research Establishment (USA, defunct)
SLAP	Standard Light Antarctic Precipitation (isotopic standard)
SLR	Sea Level Rise
SMCL	Secondary Maximum Containment Level
SMD	Soil Moisture Deficit
SMOW	Standard Mean Ocean Water
SOR	Successive Over Relaxation
SP	Self Potential = Spontaneous Potential
SPAC	Soil-Plant-Atmosphere Continuum
SPCZ	South Pacific Convergence Zone
SPOT	Satellite Probatoire de l'Observation de la Terre
SPS	Safeguard Policy Statement
SR	Rubber modified polystyrene
SRB	Sulphate Reducing Bacteria
SRI	System of Rice Intensification
SS	Suspended Solids
SST	Sea Surface Temperature (ΔSST = incremental change in SST)
STP	(1) Standard Temperature and Pressure (= n.t.p)
	(2) Sewage Treatment Plant

Dictionary of Hydrology and Water Resources

SUH	Synthetic Unit Hydrograph
SWL	Static Water Level
TA	Technical Assistance
TBA	Trend-Based Analysis
TCA	Tricarboxylic acid
TCM	Total Catchment Management
TD	Total Depth
TDI	Tolerable Daily Intake (or Ingestion)
TDR	Time Domain Reflectometry
TDS	Total Dissolved Solids (units of mg.l-1)
TEM	Transient Electro-Magnetic
TGA	Thermo-Gravimetric Analysis (= DTG)
THM	Trihalomethane(s)
TIROS	Television and Infra-Red Observation Satellite
TISAB	Total Ionic Strength Adjustment Buffer
TKN	Total Kjeldahl Nitrogen
TL	(1) Thermo-Luminescence
	(2) Team Leader (of project or program)
TM	Thematic Mapper
TMDL	Total Maximum Daily Load (USA)
tntc	Too numerous to count. (bacterial colonies on an agar plate)
ToC	Top of Casing
TOC	Total Organic Carbon
ToR	'Terms of Reference'
ToT	Training of Trainers
TPC	Total Plate Count (microbiology)
TPH	Total Petroleum Hydrocarbons
TSE	Treated Sewage Effluent
TSI	(1) Tropical Storm Intensity
	(2) Total Solar Irradiance
TSS	Total Soluble Salts; Total Suspended Salts (*not the same!*)
TUH	Time Unit Hydrograph
TVA	Tennessee Valley Authority
TVC	Thermal Vapour Compression (distillation)
TWS	Terrestrial Water Storage
UNDP	United Nations Development Programme

UNEP	United Nations Environment Programme
UNESCO	United Nations Educational, Scientific and Cultural Organization
UNFCCC	United Nations Framework Convention on Climate Change
UNOPS	United Nations Office for Project Services
USBR	United States Bureau of Reclamation
USEPA	(or just EPA) United States Environmental Protection Agency (standards)
USGS	United States Geological Survey
USLE	Universal Soil Loss Equation
UST	Underground Storage Tank
UTM	Universal Transverse Mercator (map projection)
UV	Ultra-Violet
UWM	Upper Water Mass (of a stratified water body)
VCH	Volatile Chlorinated Hydrocarbon(s)
VDS	Visual Display System
VES	Vertical Electrical Sounding
VHO	Volatile Halogenated Organics
VOC	Volatile Organic Compounds (or Chemicals)
VPD	Vapour Pressure Deficit
VSB	Vegetated Submerged Bed
WASH	Water, Sanitation and Hygiene. (program, q.v.)
WB	World Bank (a major donor of water projects)
WDN/S	Water Distribution Network/System
WDSC	Water Dispersible Sand and Clay
WFD	Water Framework Directive (of the EU)
WHO	World Health Organization
WML	World Meteoric Line
WMO	World Meteorological Organization
WPCF	Water Pollution Control Federation
WPZ	Wellfield Protection Zone
WRC	Water Resources Centre (UK)
WRC	Water Resources Council (several countries)
WRI	(1) Water Resources Institute, (2) Water-Rock Interaction (occasional symposia)
wt	Water Table

WTP	Willingness to Pay
WUA	Water Users Association
WWF	World Waster Forum
WWTP	Waste-Water Treatment Plant
XRD	X-Ray Diffraction
XRF	X-Ray Fluorescence
zfp	Zero Flux Plane
ZHD	Zero Head Differential
zpc	Zero Point of Charge

APPENDIX

Appendix   Selected Water Sector Software Packages, as of 2018.

Note: Climate-change models not included

Not a week goes by without another water-sector model appearing on the market. A definitive list is impossible, so the following is intended merely as a guide to some of the more widely used and 'industry standard' offerings. The 'latest version' is the most recent reference that was available at the time of this review (Jan 2014), but of course, most of these packages are updated regularly, and many are completely superceded within a few years.

Application Name	Latest version	Acronym or Purpose
AQTESOLV 4.5	2016?	Advanced Aquifer Test Analysis: pump, slug and constant head
AquaCrop 6	2018	Focus on climate change, CO2. Crop yield responses to water
AQUIFEM-N	2006	2D and quasi 3D Groundwater Model including transport flow
CERES 1.14	2018	Cereal growth, water & yield, simulation
CROPGRO	2017	Grain legume growth, water & yield simulation.
CropSyst 4	2018	Multi-year cropping system simulation model
DAMBRK	1988	Dam break flood forecasting model. Discharge and travel time.
DRAINS	2018	Design and Analysis of Urban Stormwater Drainage Systems
EPIC	2006	Environmental policy integrated climate. Soil erosion, productivity, leaching.

Application Name	Latest version	Acronym or Purpose
FEFLOW 6.2	2017	Finite Element subsurface FLOW system', groundwater flow in porous & fractured media.
Flo-2D	2013	Flood routing, channel and overland.
FLONET/TR2	2013	2D steady state groundwater, contaminant transport and residence time.
FLOWPATH II	2007	2D Groundwater flow, remediation, and wellhead protection model
GWLF	2018	Generalized Watershewd Loading Function. Sediment and nutrient loading.
HEC-1	2013	Basin model for rainfall-surface runoff simulation.
HEC-Ras 5.0.4	2018	1D & 2D steady and unsteady river flow simulation (Replaced HEC-2)
HEC-ResSim 3.1	2013	Reservoir simulation and management model
HYDATA 4.2	2004	Hydrologic data base
HydroCAD	2013	Modeling stormwater runoff. Design of stormwater management systems.
HYDROGEOCHEM 5	2018	Hydrologic Transport and Mixed Equilibrium Reactions (Saturated-Unsaturated)
Hydrogeosphere	2012	FE surface and groundwater model for water, solute and heat flow.
HYRROM	2004	Catchment Rainfall Runoff Model (Q from P & E)
ILSAX	1993	Simplified runoff & urban stormwater drainage design model
INCA 3	2014	Integrated Catchment Model. Pollutant dynamics
IQQM	2016	Integrated Quantity and Quality Model
ISIS 2D 3.7	undated	2D Flood modelling

Dictionary of Hydrology and Water Resources

Application Name	Latest version	Acronym or Purpose
ISIS-Tuflow	2013	Flood and Tide Simulation
IWFM	2016	Integrated water flow model
MAGIC	2010	Modelling Acidification of Groundwater in Catchments
MIKE 21	2012	Simulates flows, waves and sediment fluxes in coastal, estuary, & river sites
MIKE BASIN	2017	GIS-based basin simulation model.
MIKE FLOOD	2012	Urban, coastal and river flood simulation model.
MIKE River	2017	A river-basin model for flow, water level, sediment, flood-plain management.
MIKE SHE	2017	Dynamic catchment model, including surface and groundwater, soil & crops.
MINEQL+	2015	Aqeous species and interactive environment model
MINTEQA2	2011	Models Aqueous Species and surface adsorption
MODFLOW P	2018	3D groundwater flow model, parameterized estimation version of MODFLOW 6
MODFLOW-surfact	2018	3D Fin-diff groundwater flow, with surface water interaction & pollutant transport
MODHMS	2011	Conjunctive surface/subsurface framework. Intervaces with MODFLOW
MODPATH	2018	3D solute pathway / particle-tracking post-processing model for MODFLOW
MODSIM	2017	Generalized river basin decision support system
NASA SWOT	2017	Global Survey of Earth's water including lakes, wetlands and rivers
PHREEQC	2013	pH-Redox solute transport, mixing and speciation. Hydrochemistry, geochemistry.
RAINS	2006	Regional Air Pollution Information and Simulation model

Application Name	Latest version	Acronym or Purpose
RIBASIM 7	2012	River basin simulation model
RiverWare 7.2	2018	Water accounting, water right,s and multi purpose reservoir optimization.
RORB 6.1.8	2012	Rural and urban runoff routing package
SOBEK 2D	2018	Flood behavioue and overland flow hydraulic model.
SOLMNEQ	1988	Solute Mineral Equilibrium (Multiphase up to 350°C)
SUTRA	1997	2D and 3D Groundwater model
SWAGMAN	2013	Salt, Water and Groundwater Management and on-line water use efficiency.
SWAMP	2010	Stormwater management programme & DSS. Runs with ArcGIS
SWAP/WOFOST	2012	Analysis & simulation of the growth and production of annual field crops.
SWAT	2018	Soil and Water Assessment Tool. Management of water sediment, pollutants.
SWIM 3	2012	Solid waste Integrated management in a river basin
SWMM	2017	Stormwater Management Model
TUFLOW	2014	1D or 2D Simulation of flood and tidal flow.
UM3	2013	USEPA plume model. Mainly marine but can be lake or broad river.
VS2DT 3.2	2006	Water and solute transport model in Variable Saturation
WATEQ4F	2011	Hydrochemical speciation.
WaterCress	2012	Water - Community Resource Evaluation and Simulation System. Stormwater.
WaterWare	2010	Modular water database and web interface

Application Name	Latest version	Acronym or Purpose
WBM	2013	Water Balance Model. Irrigation, reservoir, regional water balance.
WBNM 2017	2017	Watershed Bounded Network Model (Unisearch, U.K.)
WEAP	2017	Water Evaluation and Planning System, water demand, supply, runoff etc
WHAM7	2011	Windermere Humic Aqueous Model for organic equilibrium speciation
WMS	2013	Watershed Modelling System, hydrology, floodplain, storm-drain
WTAQ 2.1	2017	Drawdown -test simulation for confined and water tablke aquifers.
XP-RAFTS	2013	Non-linear Runoff Analysis and Flow Training System, urban & rural.